THIS WILL KILL YOU

THIS WILL KILL YOU

A GUIDE TO THE WAYS IN WHICH WE GO

HP Newquist

Rich Maloof

Illustrations by Jim Shinnick

St. Martin's Griffin New York

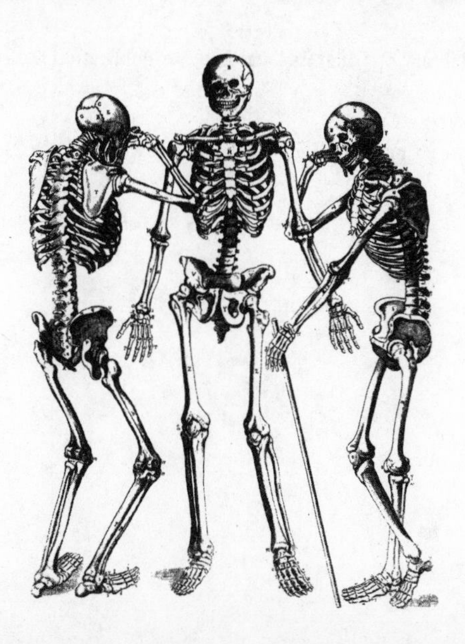

The authors have created this book with the intention of presenting information on forms of death. In no way, shape, or fashion do we recommend, encourage, or condone pursuing any behavior or action described herein, for any reason whatsoever. It is our belief that the most desirable way to exit this world is in your sleep at an age that hovers around triple digits. In fact, we hope that this book will aid you in finding ways to avoid those things that endanger your well-being and help you to live long enough to have the last laugh.

 For information, address St. Martin's Press, 175 Fifth Avenue, New York, N.Y. 10010.

A Nightwatch Press Book

www.stmartins.com

Illustrations by Jim Shinnick

Book design by Rich Arnold

Library of Congress Cataloging-in-Publication Data

Newquist, H. P. (Harvey P.)

This will kill you : a guide to the ways in which we go / H.P. Newquist and Rich Maloof.—1st ed.

p. cm.

ISBN-13: 978-0-312-54062-3

ISBN-10: 0-312-54062-0

1. Death—Causes. I. Maloof, Rich. II. Title.

RA1063.N49 2009

614.4—dc22

2009006908

First Edition: June 2009

10 9 8 7 6 5 4 3 2 1

HP NEWQUIST:

To Trini, Madeline, and Katherine . . . for always taking the time to laugh at the funniest things in life

---

RICH MALOOF:

For my family, who shines light into every dark corner

CONTENTS

ACKNOWLEDGMENTS

Thank you to Sarah Lumnah, of St. Martin's Press, whose dark sense of humor is matched only by the brightness of her intellect. Sarah's commitment, insight, and ability to put up with us made this one of the most enjoyable projects either of us has ever worked on. And we're sorry to end that last sentence with a preposition.

Thanks to artist Jim Shinnick, whose images and icons bring these pages to life. Jim's unique view of the world, which can be glimpsed between the lines of his black and white illustrations, was essential to the creation of this book.

Thanks to all the people at St. Martin's who made this a better book while maintaining our original vision for it. Gregg Kulick created the great cover art and Rich Arnold designed the entire interior. Much thanks also to Matthew Shear, Elizabeth Curione, Eric Gladstone, Joe Goldschein, and John Karle.

And thanks to Dr. Michael Collins, for reviewing our final draft and offering his generous comments.

And finally, thanks to Peter Fitzpatrick, M.D., and Bill McGuinness, who stepped outside of their respectable chosen professions to write the insightful foreword and afterword, respectively. Their contributions make outstanding bookends to our work.

HP NEWQUIST WISHES TO THANK . . .

Thanks first to Rich Maloof, whose ability to engage in late night conversations about everything from music and medicine to morgues and meningitis has made him a stellar writing partner and invaluable friend. He is one of the few people anywhere who can cross from the bright side to the dark side and back again . . . while making both of them laugh-out-loud funny.

A world of thanks to Jim Shinnick, the finest art director in America; the Carlsons; the Cemo family; Tucker Greco and family; Thomas Werge; Ken Wright; Al Mowrer; Dan Collins; John Kunkel (RIP); the staff of the Lou Dobbs radio show; Michael and Barbara Johnson; frequent coconspirator and fellow author Pete Prown; José Cara, Senior Medical Director at Pfizer Global Pharmaceuticals, for encouraging my work on this book and giving it a professional going over; and Dr. J. Michael McWhorter—in memoriam—who spurred my interest in all things medical and surgical.

Thanks to my parents, brothers, sisters, and their families, who are as funny a group of humans as any writer could ever hope to be related to. And eternal gratitude to Trini, Madeline, and Katherine, without whom this—and all my endeavors—would just be words on a page.

Finally, I want to single out the members of Skid Row in Pangborn Hall at the University of Notre Dame. There are too many of you to be listed here, but you know who you are, and I thank you for decades of friendship and twisted humor, especially those who have been especially instrumental in keeping the Skidder spirit alive: Mike Johnson, Bill Leary, Bill Brahos, Terry Johnson, Peter Fitzpatrick, and Bill McGuinness. After all these years, you guys still kill me.

RICH MALOOF WISHES TO THANK . . .

Harvey Newquist, friend and *Nightwatch Press* writing partner, who dreamed this project in the first place and saw it through to the last. This is a guy who can speak knowledgeably about Chinese export tariffs, argue the virtues of Godflesh's guitar work, and then wrap with a description of python anatomy. For knowing the facts and finding the laughs every time, here's to Harvey.

Thanks to Dick Larimer, for encouraging and largely underwriting my plunge into health writing. Thanks also to Peter Luyckx, Jeri Condit, Susannah Detlef, and all of my talented friends at MSN Health & Fitness.

Gracious thanks to the physicians, scientists, researchers, and other experts who've shared their knowledge and their genuinely Hippocratic nature with me in hundreds of interviews.

Thanks to my brother-in-law, the Reverand Vincent M. Corso, and other hospice care workers like him who give selflessly of themselves to provide comfort and peace for complete strangers. And to my sister, Christine Corso, for having the good sense to marry him.

For their ongoing, unwavering support, I'm forever grateful to my family and friends. It's a lucky man who has too many to name on a page. I can only hope my kids—Daniel, Tess, and James—are as fortunate to know good people throughout their lives who will be there till the end. Neverending thanks to my wife, Kris Aswad, who knows when to turn my eyes away from the writing and when to look away herself. Thank you finally to my parents, Mitchell and Teresa Maloof, who every day provide a model for a life well lived.

FOREWORD

BY PETER M. FITZPATRICK, M.D.

ASSISTANT PROFESSOR OF MEDICINE,

MAYO MEDICAL SCHOOL, JACKSONVILLE, FLORIDA

Man has feared and been fascinated by death since the earliest of times. Many of the oldest artifacts from the ancient world were related to death: amulets to protect the dead, tombs to encase the dead, inscriptions to honor and glorify the dead. Poems have been written, songs composed, and epitaphs chiseled. But what do we really know about death?

When I was in medical school we didn't talk about death. There was too much to learn about what keeps people alive. As a resident, I was armed with weapons to keep death at bay: advanced Cardiac Life Support techniques, monitors of all sorts, and a seemingly endless list of drugs and medications. Yet, death would come sometimes. The first time I attended someone's death, I remember the uncomfortable feeling that I had just witnessed something that was very profound . . . yet common. That feeling has never gone away.

Death can be viewed as an event or a process. And certainly, in the broadest sense, death is the cessation of life. But when does death actually occur? Even in those instances where death would seem easy to define, there are still questions. For instance, a decapitation: One presumes the person dead when the blade slices though all the structures of the neck and the head falls to the ground. But how long does it take a severed head to realize that it is no longer attached to the rest of the body before it slides into a state of unconsciousness? Is that when death occurs? If the heart continues to beat and pump blood for several minutes—when it stops, is that death? A shark's heart will beat for several hours after being totally removed from its body. When did the shark die?

In the early 1800s, there was a general fear of being buried alive due to the fact that the medical definition of death was not precise. Several patents were obtained in that time for "safety coffins," which would allow a person who was not yet dead, but buried, to give a signal by means of pulling a cord that was attached to a bell to alert a watchman. Thus, it is said, was the origin of the expression "saved by the bell." The stethoscope, that universal image associated with the healing arts, was developed to help a doctor listen for a heartbeat. If nothing was heard for two to five minutes, it was time to officially declare the person dead.

In more modern times, medicine has adopted a "brain death" definition of death. This was felt necessary due to medical technologies that now allow us to keep hearts beating, lungs inflated with air, and blood cleansed with machines for weeks, months, or years. In many ways, the presence of brain waves that indicate higher brain function are the modern-day version of the bell on a cord—the person is not yet dead but entombed by machinery.

This definition has lead to ethical, philosophical, and spiritual debate about the persistent vegetative state, when higher brain functions are absent but lower functions persist. Is a person truly alive when only the most basic functions are preserved but the personality and intellect are gone? The unknowns even get us into the realm of asking where the human soul resides.

Defining death as a process is even more difficult. Programmed cell death, known as apoptosis, begins during embryonic development and continues all through what we call "life." Since this is the case, do we start dying at the moment of conception? After all, the obsolescence of the body is a rather predictable process. For instance, we lose about 7 percent of kidney function for each decade after age twenty. But the orderly processes of life can be derailed by various forces: external (trauma and exposure); internal (infections, cancer, smoking, and alcohol); or by chronic and degenerative disease. These are the things that will kill you.

Yet, as medical science learns to control or treat one chronic disease, it seems to open the door to other, previously unforeseen consequences. For example, diabetes can be treated effectively and save diabetics from early death caused by uncontrolled blood sugars. Unfortunately, many diabetics develop kidney failure as they live longer. Of course, once kidney failure sets in, they can be kept alive with dialysis treatments. Then again, the cardiovascular complications of diabetes continue on, so many diabetic dialysis patients die from heart attacks. One thing we know for sure is that cheating death is tricky business.

This book explores many of the countless ways we can die—some rather exotic and some mundane. I hope that it will answer some of the questions you might have about death and dying. The truth is that it will likely raise even more questions than it answers. That's to be expected, though, because death—for everyone, even doctors—truly is the great unknown.

THIS WILL KILL YOU

INTRODUCTION

We are all, every one of us, going to be laid to rest at some point in the future. There is no escaping it. A day will come when "living life to the fullest" is no longer an option—or even a possibility. We will be dead.

None of us likes that thought. The whole idea of not being around to enjoy life really unnerves human beings. For good reason: we'll miss the good times, we'll miss out on time with our family and our friends, we'll never get to finish all that we set out to do. We'll be done. Over and out.

While such concerns are worth discussing at great length, that's not going to happen in this book. We're not here to talk about the metaphysics of dying or to ponder the mysteries of the Great Beyond. We're here to talk about the way that life ends; in fact, a lot of the ways that life ends. While we all want to go peacefully in our sleep rather than screaming hysterically as we attempt to outrun a grizzly bear, the fact is that few of us really know how we're going to go unless we plan it ourselves or a state correctional facility is arranging a special event for us.

There are lots of ways to go. Some of them are incredibly unpleasant, some not so bad, and some are actually kind of funny as long as they're not happening to you.

Most of us are familiar with the events that cause our always untimely demise—diseases, accidents, lethal force—but we typically don't know why such things make us die. For example, what really happens to the human body when you get hit by lightning? What does the dreaded Ebola virus really do to your insides? Why will a speck of ricin the size of this dot (·) kill you, yet you can survive having your entire arm hacked off? Why will four shots of whiskey make you happy but forty shots make you dead?

It all comes down to the moment when your body has had more than it can stand, and you take the big leap to the hereafter. And since you're definitely going to go, you might as well know what happens to you. That's why we wrote this book.

The human body is the single most complex organism in the known universe. When you consider the functions of your brain, heart, and lungs, it is amazing that your organs work day in and day out as well as they do. If your body were manufactured by an American car company, for instance, you'd have been dead decades ago—no matter how old you are now.

The sheer number of moving parts inside you, all operating in harmony, is astronomical. And the statistics on what's going on in your body at any given moment are, to put it mildly, death-defying.

Your lungs breathe in approximately six liters, or 1.5 gallons, of air every minute, about 20 percent of which is oxygen. That's equal to nearly 9,000 liters of air a day.

From your lungs, oxygen is transferred to your blood. There are about seven pints of blood in the average adult, or just under a gallon. Oxygen-rich blood is pumped through your body by the continuous beating of your heart. The human heart beats roughly seventy times per minute, during which it pumps a gallon of blood through the nearly 100,000 miles of arteries, veins, and capillaries in the circulatory system.

Your heart, should it last as long as you hope it will, beats more than two billion times in your lifetime. Each day, some 100,000 heartbeats pump the equivalent of 2,000 gallons of blood to every corner of your body, and we do mean every corner. Try pricking any area of your body with a needle, and you're sure to draw blood.

The largest percentage of fresh blood, one-fifth, goes straight to your brain to deliver oxygen as the fuel that keeps your brain functioning. Every second of every day, the more than 100 billion neurons in your brain are creating thoughts, controlling your organs, and helping you to figure out how to stay alive. These neurons are connected to each other via an estimated 100 trillion synaptic connections. The firing of these synapses controls your heart rate, your breathing, and the operation of your internal organs, not to mention the storage and processing of your thoughts and memories.

The reason that all of this high-school-level biology information is so important is that when something interferes with these processes, you are in serious danger of dying.

What, then, is death? Avoiding any religious and philosophical musings, as well as jokes about marriage and bad jobs, there is an actual definition of death. It is found in the Uniform Determination of Death Act of 1980. Created by the National Conference of Commissioners on Uniform State Laws, and approved by frequent death-profiteers the American Medical Association and the American Bar Association, this act puts it all very simply:

> An individual who has sustained either (1) irreversible cessation of circulatory and respiratory functions, or (2) irreversible cessation of all functions of the entire brain, including the brain stem, is dead.

For a definition created by committee, this is astonishingly impressive for its brevity. It basically says there are only three ways you can be classified as dead: your heart stops, your lungs stop, or your brain stops. Permanently.

Medical purists may argue that all death is brain death, since a nonfunctioning brain can no longer command internal organs to continue working. To them, it follows logically that the breakdown of the respiratory system keeps oxygen from feeding the brain, and that lack of blood flow achieves the same sordid result. It's all about the brain . . . and we sympathize. However, in recent years, the use of "brain dead" has referred to different elements of the brain: the cessation of thinking or consciousness, and the termination of control over bodily functions. This has caused no end of confusion among the general public.

The fact is that you can survive, albeit not pleasantly, without higher level brain functions. But once the parts of the brain that control autonomous body functions—like the beating of your heart, the expansion and contraction of your lungs, your digestive system, the reproduction of cells, and the production of hormones—shut down, you are toast.

It must be noted that if you have no brain, you die. Other organs can be replaced, transplanted, or kept alive with a machine. The brain is the one organ on the menu that has no substitute of any kind. Thus, the purists leading the "all death is brain death" contingent can hold their heads up high even in light of the Uniform Determination of Death Act.

Arguments of medical mundanity aside, everything in this book, regardless of how, where, and why they take place, will kill you in one of three basic ways that fit the definition: they will stop your blood from flowing, they will stop your body from sucking in oxygen, or they will stop your brain from performing its essential functions. These are the three simple ways in which things will kill you.

It is the diverse routes taken by diseases, accidents, and aging to achieve these three things that make death a particularly interesting, albeit ghastly, endeavor. In the following pages, you'll find out how sushi can short-circuit your brain, how constricting snakes can force your heart to burst, and how a botox injection can paralyze your lungs.

You will become acquainted with the obvious and not-so-obvious modes in which you might be moved to the mortuary. Animals, plants, bacteria, viruses, fungi, chemicals, extreme temperatures, fast-moving objects, sudden impacts, out-of-control cells, air, water, and even Acts of God are all featured here in their myriad lethal forms.

As the authors of what you are about to read, we know that we have to go someday, too. Hopefully it won't be until long after you've purchased this book and learned a little bit from it. Our wish is to go peacefully and painlessly in the middle of a really good dream, at least fifty years from now. The hard part for us—and for you, for the president of the United States, for the Dalai Lama, for psychics and fortune-tellers everywhere—is that one never knows.

One final note: The phrase, "This will kill you" has long been used by storytellers and comics to preface what they believe is going to be a classic bit of good humor. Its origins go back to the theater, where it was considered bad luck to wish someone good luck (hence phrases like "break a leg" and "knock 'em dead"). We'd like to think that we've incorporated some of that sentiment in the pages that follow. While reading won't literally kill you, we do hope that this book will kill you. Figuratively speaking, of course.

TWO-MINUTE MED SCHOOL

In the pages to follow, you're going to find out a lot of things about your body that you probably didn't know. Much of it is going to involve strange words with lots of syllables and vowels in weird places. Unless you went to medical school, terms like infarction and hypovolemia are likely to make you think of countries in the former Soviet Union.

To help you sort out these terms, and get a glimpse into the workings of your anatomy, we're adding this special section as a bonus. We call it "Two-Minute Med School," and it's included here free of charge (and there's nothing in real medical schools that is ever free of charge). After reading "Two-Minute Med School," you will have the same command of scientific and medical terms normally reserved for emergency-room surgeons—just like those played on TV.

Once you're familiar with these terms, you'll be able to use them at parties and other social gatherings, not to mention in private conversation with your physician. Go ahead, impress everyone you know by noting that an inflammation is not the same thing as an infection. Amaze your friends by pointing out the difference between a heart attack and cardiac arrest. And if you happen to be having one, you can tell the ER guy what's happening and just maybe speed things up at the hospital.

If nothing else, refer back here when you see a word you don't know, just like the doctors do with all those books on the shelves in their offices. Oh yeah, they're looking this stuff up more often than you'd like to think.

Antivenin: an antidote for venom. Antivenin and antivenom are synonyms.

Anaphylactic shock: an allergic reaction to a protein, usually in the form of a bee sting, penicillin, peanut, or other entity that results in a drop in blood pressure, typically from dilated blood vessels. Also called anaphylaxis, but anaphylactic shock may be the most fun term to say in this entire book.

Asphyxiation: oxygen deprivation

Bacterium: a single-celled organism; the plural is bacteria. Dormant bacteria are spores, and spores become active under the right conditions such as those found under your skin.

Cardiac arrest: when your heart stops beating. If this happens for more than five minutes, you are going to die.

Circulatory failure, circulatory collapse: *see* Shock

Heart attack: an event that directly affects the functioning of the heart. Think of it as something that attacks the heart, like a restricted supply of blood; some sort of interruption. A heart attack, if left untreated, may lead to cardiac arrest, which is the actual stopping of the heart. That's when things get fatal. Cardiac arrest is the stage at which the heart ceases to work, and this will kill you.

Hemorrhage: profuse or abnormal bleeding. This often occurs when blood vessels are ruptured or severed. For instance, a cerebral hemorrhage occurs when blood flows into brain tissue from broken blood vessels.

Hypovolemic shock: a state of shock caused by reduced blood volume. This is usually caused by severe bleeding and the resultant blood loss. When too little blood enters the heart, too little blood is pumped to the brain and the rest of the body.

Hypoxia: a low level of oxygen in the tissues of the body, usually resulting from compromised blood flow or low oxygen levels in the blood.

Infarction: the death of tissue when blood supply is cut off to that tissue.

Infection: what happens to your body when it is invaded by disease-causing microorganisms such as bacteria and viruses. The growth of these microorganisms as they feed off of your body creates an infection.

Inflammation: your body's reaction to an injury. When you are wounded, or microorganisms like viruses and bacteria attack your body, blood flow to that area increases in order to fight the culprit with white blood cells. This increased blood flow is what turns inflamed areas red. Other chemicals are released into the area to help repair the damage.

Myocardial infarction: a heart attack

Necrosis: the death of living cells and tissues. Necrosis can occur in one section of an organ or tissue while other sections remain healthy.

Neurotoxin: a toxin that destroys nerve cells. Since the largest concentration of nerves are found in the brain and spinal cord, they are extremely susceptible to neurotoxins.

Organ failure: the failure of one of your essential organs, notably your heart, lungs, kidney, liver, and brain. Multiple organ failure means that two or more of these have shut down. Multiple organ failure is also referred to as multiple organ dysfunction syndrome (MODS).

Pulmonary: relating to the lungs and the movement of blood to and from the lungs. The **respiratory** system, on the other hand, describes the lung's intake of air and involves the mouth, nose, trachea, and diaphragm.

Renal failure: the inability of your kidneys to adequately filter waste products out of your blood. Renal is from the Latin for kidneys.

Shock: a potentially deadly condition in which your blood pressure is too low to keep you alive. Shock occurs when your blood pressure drops so much that your cells don't get enough blood and therefore not enough oxygen. This prevents cells in your brain, heart, and other organs from functioning normally, leading to failure in these organs. A number of things can cause shock, but one of the most common is low blood volume, which is known as hypovolemic shock. It is critically important to note that medical shock is not the same thing as emotional shock. **Emotional shock** is a sudden psychological or emotional disturbance.

Toxin: a poisonous substance produced by living organisms. Toxins can cause disease and adverse reactions in specific areas of your body. Neurotoxins attack nerve cells, hemotoxins destroy red blood cells, and cardiotoxins damage the heart.

Trauma: physical damage that is inflicted on your body as the result of an external force. This can be a punch in the face, a car accident, a fall, a gunshot, and a host of other events that deliver an impact to your body. Trauma is an actual cause of death when the amount of force destroys organs and major blood vessels. Severe trauma is one of the primary causes of death worldwide, and the leading killer of people under the age of forty-five in the United States. Emotional trauma should not be confused with physical trauma, just as emotional shock should not be confused with physical shock.

Virus: an infectious organism that invades living cells and reproduces itself. A virus does this by releasing its DNA (deoxyribonucleic acid) or RNA (ribonucleic acid) into the host cell, which makes new copies of the virus. The host cell then dies and releases new viruses, which find new hosts. Viruses are smaller than bacteria.

DESCRIPTION OF ICONS

Kills Per Annum: In order to give a human face—or rather, a human toll—to our entries, we've devised the following easy-to-use system of icons. It will help you immediately ascertain how many of your fellow humans are destined to be taken out of the gene pool this year by each thing that will kill you.

The arrangement is simple. Each icon stands for the toll typically inflicted by each of the pictured objects worldwide, unless otherwise stated:

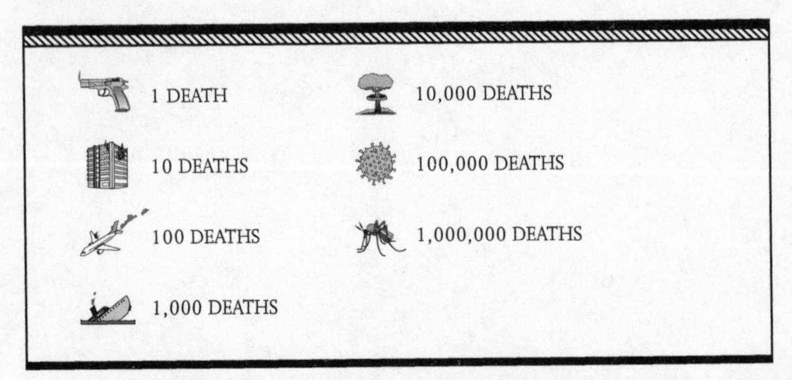

Thus, and a means twenty-one deaths, while means 5,000 deaths. More than means the Grim Reaper is having a field day. You get the picture.

Lethality: Measuring the likelihood that an encounter with this killer will make you dead. Lethality is expressed as Low, Medium, or High.

Horror Factor: Sometimes we go out easy, sometimes in a terrifying display of pain and blood and screaming. Our horror gauge indicates a point on the entire range.

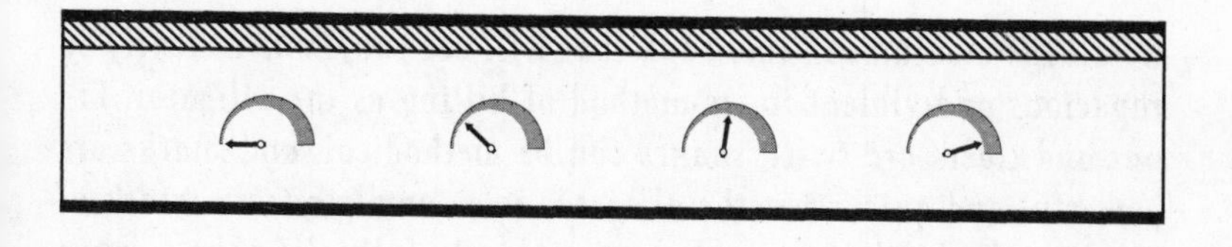

ALLIGATORS

Of all the creatures in this book that will eat you, none is nearly so rapacious and violent in its method of killing as the alligator. Lions and tigers are swift, snakes can be methodical, and sharks are powerful and quick. But the alligator is a complete train wreck of a killer, the kind of animal that would gleefully defy the Geneva Convention of animal cruelty toward humans. If there was one.

HOW IT KILLS

There is nothing pretty, clever, surgical, or swift about an alligator's mode of killing. In all cases it is brutish, violent, loud, messy, and bloody.

Alligators prefer to catch victims in their natural water habitats—lakes, rivers, canals, and bayous. Swimming makes it easy for a reptile weighing 300 pounds or more to noiselessly sneak up on its prey. And alligators, despite their size and lumbering gait, can run over land very quickly for short bursts to take down anyone inclined to run away.

Once in striking distance, the alligator

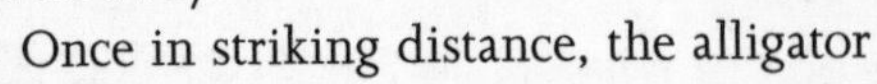

clamps its jaws over any available body protrusion, such as an arm, leg, or head. Alligator jaws have the greatest force of any land animal, measuring in at a bone-crushing 3,000 pounds per square inch (PSI). Thus, pulling your arm, leg, or head away from an alligator's mouth intact is not going to happen. In addition, its feet are clawed much like that of a velociraptor, and can rip through muscle and bone like a razor through cloth.

Once it has you securely in mouth, the alligator's favorite method of killing is to drag you—kicking, screaming, thrashing—deep into the water. This serves the purpose of speeding up the death process by drowning you. Alligators can stay underwater for an hour without breathing, which is about fifty-six minutes longer than you can. In the process, however, the alligator flails you about like a rag doll with epilepsy. This is called the alligator's "death roll" and is performed in order to wrench the aforementioned limb out of the appropriate socket (shoulder, hip, etc.) while you concentrate on trying to breathe.

Once your body stops moving, the alligator drags your corpse back to dry land where it proceeds to tear large sections away from whatever remains. For anyone who is not quite dead, and has only passed out from near-asphyxiation, this is about as close to living Hell as it gets right here on Earth.

KNOWN BY SCIENCE AS:

There are only two true alligators: the American alligator (*Alligator mississippiensis*), and the Chinese alligator (*Alligator sinensis*). Crocodiles, found in Africa and Australia, are not-too-distant relatives as both crocs and gators belong to the order *Crocodilia*.

MEDICAL CAUSE OF DEATH

Asphyxiation; organ failure; blood loss

TIME TO KILL

Variable, because it depends on how long you can hold your breath or how long you can survive with a missing limb and no bandages. Our estimate is you have about four to six minutes to get your affairs in order.

HIGHEST RISK

Athletes in Florida—especially golfers, swimmers, and joggers; gator hunters in Louisiana swamps

LETHALITY

Low, about 15 percent. Between six and ten attacks are recorded every year, but only one or two fatalities. Many of the attacks are caused by people bothering alligators that are basking along canal banks or sunning in open areas.

KILLS PER ANNUM

One or two. Not very many, but they sure do make the headlines when they happen, due to the victim fragments that are occasionally left behind.

HISTORIC DEATH TOLL

Hundreds since the settlement of the Americas. For all its power and grim determination, the alligator has not compiled the kinds of killing numbers posted by, say, tigers and snakes.

NOTABLE VICTIMS

Annemarie Campbell, Judy Cooper, and Yovy Suarez Jimenez were all killed in Florida in the space of one week in May 2006. In the last six decades, only seventeen people have been killed in Florida by alligators. That's one every three years or so. Yet in that one week, these three women were killed in broad daylight and in populated areas. All three became part of the revered circle of life, as if the gators were on some kind of human-protein South Beach Diet.

HORROR FACTOR: 11, ON A SCALE OF 10

Having a creature that looks like a dinosaur drag you to the bottom of the swamp with your leg clamped in its mouth has to give new meaning to the phrase "What a way to go."

GRIM FACTS

- Alligators have some of the strongest jaws ever crafted by natural selection, but they only exert significant force when snapping closed. Alligators have very weak musculature with which to open their mouths, and the jaws on even the biggest gators can easily be held shut with your bare hands.

- When the alligator was placed on the endangered species list in the United States in 1967, there were fewer than 300,000 in existence. Today there are an estimated two million, living everywhere from swamps to suburbs, which accounts for a recent rise in their attacks on humans.

- Crocodiles are close relatives of alligators, and feast more readily on local villagers in their native Australia and Africa than do their American counterparts. A crocodile known as Gustave is believed to have killed more than 200 villagers in Burundi since the 1980s—and is still at it.

ALZHEIMER'S DISEASE

Notorious for its ability to destroy memory like it was stored on a faulty computer, this disease can just about wipe your mental hard drive clean. In the most advanced stages, you won't get a chance to reboot.

HOW IT KILLS

The aging of America's baby boomers has brought on a wave of action against diseases that take hold late in life. On the upside, that results in more motivated and better funded research on several degenerative conditions; on the downside, Alzheimer's is still well ahead of the wave. It is the most prevalent form of dementia today, though watching reality shows runs a close second.

It's hard to cure a disease when you don't understand what causes it. Ever since German neurologist Alois Alzheimer discovered abnormalities during a 1906 autopsy of a woman who had dementia, the disease has been hallmarked by tangles of neural fibers and deposits of plaque in the brain. On your teeth, plaque is unsightly; on your brain, it's a bit more of a problem. The origin of these neural plaque malformations is unknown, though risk factors include genetic disposition, head injury, exposure to toxins, and, of course, advanced age. Approximately 10 percent of those over sixty-five are believed to have the condition—and nearly half of those aged eighty-five or older. Scientists estimate some 4.5 million Americans (nearly the population of Norway) currently suffer from Alzheimer's.

Remind myself to place illustration here.
—Editor

Though AD is the nation's Number 7 killer, it's uncommon to die from direct causes. Instead, the way it compromises your cognitive functionality will bring on deadly secondary conditions. Following later-stage Alzheimer's symptoms—like forgetting how to comb your hair, failing to recognize your spouse, or having problems with speaking and comprehension—you become susceptible to other health threats. Difficulty swallowing sends food and drink into your airways and lungs, leading to pneumonia; incontinence leads to urinary tract infections, then more serious infection; falls lead to head injuries, and head injuries to strokes. You won't even find much assistance with a help-I've-fallen-and-I-can't-get-up alarm, since you probably won't remember that you bought one or that you're wearing it.

KNOWN BY SCIENCE AS:

Alzheimer's Disease (AD)

MEDICAL CAUSE OF DEATH

Amyloid plaques and neurofibrillary tangles in the brain leading to secondary conditions such as pneumonia, sepsis, and stroke

TIME TO KILL

Average of eight years after diagnosis; sometimes as long as twenty years

HIGHEST RISK

Susceptibility appears highest among people who are over eighty-five; have sustained head injuries; have immediate relatives with Alzheimer's; are inactive or physically unfit

LETHALITY

High, although prolonged

KILLS PER ANNUM

Approximately 72,000 in the United States

HISTORIC DEATH TOLL

Deaths attributed to Alzheimer's rose 25 percent between 1998 and 1999, when related secondary conditions were first tabulated as AD fatalities. Outside of the world of statistics, there have been dramatic improvements in the survivability due to improvements in diagnosis, awareness in the medical community, and other factors.

NOTABLE VICTIMS

Physicians believe the head injury Ronald Reagan sustained when thrown from a horse in 1989 hastened the onset of Alzheimer's, which he was diagnosed with five years later. Media coverage of Reagan's decline helped raise awareness of the disease. Other famous victims include Rita Hayworth, Charlton Heston, Iris Murdoch, and James Doohan (Scotty on *Star Trek*).

HORROR FACTOR: 8

Alzheimer's earns a high horror factor since the suffering it causes a patient has to be added to the dread and emotional burden on loved ones.

GRIM FACTS

- A study at the University of Bordeaux found that people who drank up to two glasses of wine per day had slightly diminished risk of dementia; still, no responsible medical professional would recommend wine consumption to prevent Alzheimer's.

- An association has been made between less education and increased incidence of Alzheimer's. Researchers believe increased use of one's intellect may reduce risk, perhaps because an active brain creates more synapses.

- Contrary to long-standing theories, there's no reliable evidence that Alzheimer's is caused by exposure to aluminum.

ANAPHYLAXIS

Some allergies tickle your nose. Some even make your lips swell or give you a case of the runs. But when a bee sting—or a simple sip of milk—leaves you gasping for air, you've got yourself a genuinely deadly allergic response known as anaphylactic shock.

HOW IT KILLS

Our immune systems are still evolving, and because of that sometimes they just plain goof. The essence of an allergic reaction is having your body react to something as if it's dangerous when it really isn't. Of course, if you try telling that to someone in the emergency room after they've innocently enjoyed some walnuts, you're apt to get a dirty look. If they can still open their eyes.

The problem—and for 30,000 Americans every year, it is a very big problem—is that the proteins in some foods, chemicals, and insect venoms can cause certain cells (called mast cells) to explode and release histamines. This wreaks havoc on the entire system, and can lead to a serious, potentially fatal allergic reaction known as anaphylaxis. The most common

culprits are foods like fish, shellfish, tree nuts, eggs, and milk; stinging insects like yellow jackets, wasps, and fire ants; and medications like antibiotics and antiseizure drugs. Products containing latex, too, can cause anaphylaxis, putting hundreds of surgical patients at risk when a surgeon's glove comes in contact with their soft, pink innards.

So imagine you've just had your first bite of a lovely salad sprinkled with nuts and dressed with egg whites. First your mouth starts to tingle, then itch. You detect a metallic taste. Your curiosity turns to fear as your tongue swells, your body begins to feel warm all over, and your skin breaks out in an agonizing rash. Within minutes, or even seconds, your stomach cramps—and then panic sets in as you realize it's increasingly difficult to draw a breath.

Anaphylaxis can be devastating to the respiratory system, tightening your airway and squeezing your chest with a pain that can mimic a heart attack. In fact, your heart is potentially in trouble since anaphylaxis not only narrows your bronchial tubes but causes your blood pressure to plummet. Shocked by the inability to draw air and/or maintain blood pressure, your body gives up and you begin to pass out. As the world goes black, you realize the error of your ways and wish you'd just ordered the chicken fingers.

KNOWN BY SCIENCE AS:

Acute anaphylaxis

MEDICAL CAUSE OF DEATH

Respiratory shock; hypotension

TIME TO KILL

Minutes. In a biphasic (two-phase) reaction, symptoms subside for a few hours only to return with deadly consequence.

HIGHEST RISK

Young people aged ten to nineteen who have both a food allergy and asthma are in the group with the highest fatality rate.

LETHALITY

Low, except for those with asthma or previous anaphylactic responses. Anaphylaxis is believed to affect as much as 15 percent of the U.S. population.

KILLS PER ANNUM

Approximately 1,500 from all types of anaphylaxis. About 150 people die every year from food allergies, another 40 from insect stings. Penicillin, the wonder drug of the twentieth century, doubles as a lethal allergen, leading to some 400 deaths every year.

HISTORIC DEATH TOLL

Undetermined, though anaphylactic reactions are on the increase worldwide. In the ten-year period between 1994 and 2004, the rate of hospital admissions jumped by 8.8 percent per year. However, increased awareness and the availability of epinephrine and antihistamines as emergency treatments has held the death toll relatively steady.

NOTABLE VICTIM

In 2001, a nineteen-year-old from South Carolina who had mold allergies died after eating pancakes made from a dated flour mix. It was the last jack he flapped.

HORROR FACTOR: 6

Meeting your fate from eating a simple meal—or, worse, by taking medications that are supposed to make you healthy—is a shocking and cruel twist.

GRIM FACTS

- Ten million Americans are allergic to their own pets. Reactions to a cat or dog's saliva, dander, or urine put thousands in the hospital every year.

- Most children outgrow allergies to milk, eggs, or wheat by the age of ten.

- It wasn't long ago that a school cafeteria could be filled with kids eating peanut butter sandwiches, and no one keeled over dead into a PB&J. Researchers aren't sure why peanut allergies are on the rise, but the number of reactions doubled in just five years between 1997 and 2002.

Horsing Around

Kenneth Pinyan of Enumclaw, Washington, died on July 2, 2005, shortly after his colon was perforated during intercourse with an Arabian stallion. Though Pinyan regularly engaged in this bizarre activity, and even videotaped it, on this particular occasion he contracted peritonitis (inflammation of the lining of the abdominal cavity), which ultimately killed him. Pinyan was the subject of the 2007 documentary *Zoo.*

ANTHRAX

Anthrax was once an obscure disease feared by cattle ranchers in the Wild West. Thanks to terrorism, it's now feared by everyone from media celebrities to postal workers.

HOW IT KILLS

Anthrax is a disease that is usually found in farm animals—called grazing herbivores by science—but it can be transferred to humans via direct contact with these animals. There are three types of anthrax: inhalation anthrax, which results from breathing in anthrax spores; gastronintestinal anthrax, which comes from eating an animal that had anthrax; and cutaneous anthrax, which you get when a spore enters a wound on your skin.

We're going to presume you've inhaled anthrax. It may take a while to show any symptoms, but when they occur, things happen quickly. First, you'll feel like you have nothing more than a cold and all that accompanies it, like a fever, sore throat, and fatigue. After a few more days, you'll have trouble breathing. This is because you'll have swelling in your lungs from where the initial spores lodged themselves. Your blood will start to get less and less of the oxygen it needs.

The bacteria then begins to destroy the lung tissue and spreads to your lymph glands via the bloodstream (the spores will continue to germinate in your blood). Swollen lymph glands will put pressure on your trachea and add to the difficulty in breathing. From there, anthrax continues on its way to your brain, causing bleeding in the meninges—the thin layer of tissue that surrounds and protects your brain from infection. At the same time you'll experience hemorrhagic mediastinitis, a nice name for bleeding that occurs in the space between the lungs. A coma may follow, but death from lots of bleeding inside your body—and a lack of air—will be what finally does you in.

KNOWN BY SCIENCE AS:

The bacterium *Bacillus anthracis* causes anthrax. It comes from the Greek word for coal—*anthrakis*—because anthrax in some forms causes black lesions on the skin.

MEDICAL CAUSE OF DEATH

Respiratory failure; heart failure; oxygen depletion

TIME TO KILL

Anywhere from a week to a month and a half. Inhalation anthrax works slowly, but time is on its side once it's in your system.

HIGHEST RISK

Media personalities who open their own mail; workers at cattle slaughterhouses; third-world farmers; textile workers dealing with wool and goat hair

LETHALITY

High. Inhaled anthrax has a mortality rate of well over 80 percent, and even the antidote doesn't guarantee that yours odds will improve.

KILLS PER ANNUM

About one a year in the United States. More in the rest of the world, although major outbreaks kill dozens or even hundreds at a time.

HISTORIC DEATH TOLL

Hundreds of thousands. The "Black Bane" anthrax outbreak of 1613 alone is estimated to have killed more than 60,000 Europeans.

NOTABLE VICTIMS

One week after the September 11, 2001, attacks, envelopes with anthrax showed up in random mailboxes across the United States. Twenty-two people were infected, and five died, creating fears of a mailborne terrorist attack. Among those receiving the envelopes were U.S. senators and members of the media. On August 6, 2008, nearly seven years after the attacks, the FBI revealed that a government biodefense scientist, Bruce Ivins, was the culprit. Unfortunately, Ivins committed suicide five days before the announcement.

HORROR FACTOR: 2

It feels like you've caught a bad cold right up until you die. However, if you realize that you actually have anthrax, and know that you can't do anything about it, the horror factor is significantly increased.

GRIM FACTS

- Two of the ten Biblical plagues unleashed by God on the Egyptians, specifically the fifth and sixth, are thought to have been anthrax.

- An anthrax outbreak beginning on April 2, 1979, ultimately killed sixty-four people in the city of Sverdlovsk, east of Moscow. In 1992, Boris Yeltsin admitted the deaths were caused by an accidental leak of weaponized anthrax.

- News reporter Tom Brokaw was given the drug Cipro as a precaution against anthrax during the 2001 anthrax scare. There was a subsequent nationwide run on the drug.

- Anthrax spores in a dormant state may be able to exist in the soil for a century or more.

ARSENIC

As a white, tasteless powder, arsenic is historically famous for its use as a favorite poison by murderers all over the world. But the naturally occurring element takes out a quarter of a million people every year—and most of them never know what happened.

HOW IT KILLS

Arsenic is a chemical element found in nature. If you remember your chemistry, it's number 33 on the Periodic Table of the Elements. It has been used for a huge number of industrial applications over the centuries: in glass and metal production, as a pigment for paints and food, in computer chips, as a wood preservative, as a potent insecticide, and even in cosmetics and medicines. Being a naturally occurring element, it is found in many sources of drinking water.

Arsenic poisons you by bonding with enzymes that are essential for the ongoing chemical reactions in your body. When the enzymes are prevented from doing their jobs, it affects the operation of individual organs.

After you've ingested arsenic, it goes to work immediately on your gastrointestinal system where it breaks down small blood vessels, resulting in vomiting of blood as well as a form of diarrhea that is described as "bloody rice water." This is followed by severe dehydration, light-headedness, and lethargy. The arsenic moves from your stomach and intestines to your liver, spleen, kidney, and lungs, where it accumulates and disrupts normal organ functions. From there, it gets absorbed into other organs as well as your skin. It will also change the color of your skin and you will develop brown patches with spots of white so that your flesh resembles a dirt road with raindrops on it. Eventually, your kidneys and liver will fail as they become necrotic from lack of blood flow, and leakage from blood vessels will cause heart failure.

If you survive this initial onslaught, you will ultimately suffer longer term effects such as cancer of the skin, lungs, bladder, and kidney.

KNOWN BY SCIENCE AS:

Arsenic, or As, is element 33 on the Periodic Table of the Elements, and is classified as a metalloid.

MEDICAL CAUSE OF DEATH

Multiple organ failure; shock

TIME TO KILL

Several days on up to many years depending on the level of contamination. Those who get it from drinking water may not experience symptoms for more than a decade after first exposure, and arsenic-related cancers may not appear for up to forty years.

HIGHEST RISK

Third World children who drink contaminated water; factory workers; fifteenth- and sixteenth-century Italian politicians

LETHALITY

High. Arsenic poisoning does damage to many internal organs, and is not always diagnosed correctly in murders, suicides, or deaths related to drinking-water contamination. The level of poisoning is a determinant factor in the time it takes to kill.

KILLS PER ANNUM

Approximately 300,000 worldwide, primarily through drinking water contamination

HISTORIC DEATH TOLL

Millions accidentally, thousands by murder

NOTABLE VICTIMS

The powerful and scheming de Medicis of Renaissance Italy were noted for their liberal use of arsenic to dispatch political enemies. In 2007, it was discovered that several members of the clan had used arsenic to kill one of their own: Francesco de Medici, who ruled the house from 1574 until he died from arsenic poisoning on October 17, 1587.

HORROR FACTOR: 1

Because it takes a long time for all of the elements of arsenic poisoning to manifest themselves, the horror factor is mitigated by little or no sense of urgency. Unless you realize you've just ingested a fatal dose; that elevates the horror factor to the uppermost levels.

GRIM FACTS

- The chronic health problems of painters Vincent van Gogh, Paul Cézanne, and Claude Monet are believed by many to have been caused by their use of Emerald Green paint, which is made with copper and arsenic.

- The Borgias, one of the two most powerful Italian families of the Renaissance (the other being the Medicis), developed "time poisons" of measured arsenic that would kill their enemies according to a desired timetable. It is thought they may have experimented on animals to determine exact doses.

- Arsenic has long been one of the favorite poisons of mystery writers and playwrights, achieving iconic status as the form of murder preferred by the spinster Brewster sisters in the dark comedy *Arsenic and Old Lace.*

THE BENDS

Putting a crimp in scuba vacations since 1940.

HOW IT KILLS

If you've ever descended more than 10 feet in water, you know how hard it can be to hold your breath. That's not only because your body wants fresh air, but because the water is exerting ambient pressure on your entire body. Your lungs are squeezing the air trapped inside them like someone sitting on a balloon.

Descending to greater depths, air pressure steadily increases as the weight of water bears down on your body. At just 33 feet, atmospheric pressure is double the pressure at sea level and your lungs are squeezed to half their normal size. At 66 feet, it's threefold the pressure at sea level, and at 99 feet, it's fourfold. That's a lot of pressure. But, with a scuba tank on, you can still breathe because the tank's air is pressurized to the same degree, allowing for an easy flow to and from your lungs.

The air in the tank is about 80 percent nitrogen and as you breathe it, the pressurized nitrogen dissolves into the bloodstream harmlessly. From the bloodstream it passes into tissues of the pressure-compressed body, and remains there even when you ascend from the bottom.

If you ascend slowly, the nitrogen has a chance to dissolve in the body. But if you rise too fast, nitrogen bubbles get trapped in the bloodstream and body tissue. You'll be okay on the swim up, but within an hour on dry land the trapped nitrogen starts to cause deep pain in the body's joints, especially the shoulders. Legs and arms may also throb with pain, and you may become dizzy or short of breath. It can hurt like a terrible gas pain in your stomach, although if you feel it in your abdomen it's likely that you have gas bubbles in your spine. Your skin may begin rashing and shedding. Stabbing pains shooting down a limb or moving all around the body signal that your central nervous system is in trouble. The trapped nitrogen can literally be paralyzing.

When the bends kill, it's usually because the bubbles are so large or so numerous that they've blocked blood flow. Much like a clot, only made of gas instead of cells, the bubble is a type of embolism. Depending on where in the body they travel, the intravascular bubbles can cause blockage amounting to fatal gas embolisms in the heart or brain.

KNOWN BY SCIENCE AS:

Decompression sickness; Caisson disease; barotrauma

MEDICAL CAUSE OF DEATH

Gas embolism; hypoxia

TIME TO KILL

Thirty minutes to six hours

HIGHEST RISK

Dehydrated novice divers (especially asthmatics) diving below advisable depths

LETHALITY

Relatively low. For the millions of dives every year, fewer than 1,000 divers require any treatment for recompression.

KILLS PER ANNUM

Typically five per annum.

NOTABLE VICTIM

In 2007, New Zealand scuba diver James Aperahama might have survived the bends had he been airlifted directly to the decompression chamber at nearby Devonport Navy Base. Instead, he was first taken on a 48-minute ambulance drive to a hospital and then re-routed to Devonport. He suffered a brain embolism during treatment in the night, and died the next day.

HORROR FACTOR: 5

Slow and painful, but beats being eaten by a shark

GRIM FACTS

- If speargunning a mako shark sounds like fun to you, consider that big underwater game has been known to drag speargunners between depths and leave them with a case of the bends.

- Because fat retains more nitrogen than lean tissue, obese divers hold on to considerably more nitrogen than physically fit divers at the same pressure. Plus, they never look good in a wet suit.

A Cut Above

Vic Morrow, the actor best known for his role in the TV series *Combat!* was decapitated by a helicopter blade during the filming of *Twilight Zone: The Movie* on July 23, 1982. The helicopter lost control during an explosion, and dove down on Morrow, who was killed instantly along with two child actors.

BLACK WIDOW SPIDER

Spiders are the creepy crawly killers we fear, nestling under our bed sheets, or lurking in our shoes or just out of sight in some corner of the cabinet. When it comes to the black widow spider, this fear is well founded, because an excitable black widow can easily make your spouse a widow, or widower.

HOW IT KILLS

The black widow spider may be the most recognized bug in the world (spiders are not insects). Its bulbous black body and red hourglass tattoo are just as likely to cause you to break out in a sweat as the sight of a great white fin slicing through the water. It's a universal sign for "stay away or die."

Black widows stake out dark quiet places to spin their webs, which are among the strongest created by any spider. It waits patiently for an unsuspecting visitor, such as your hand or your foot, and sinks its tiny fangs into the flesh. The fangs, technically referred to as chelicerae, are less than a millimeter long, but they are sharp enough to deliver some of the most neurotoxic venom in the world right into your bloodstream.

This venom, called alpha-latrotoxin, is gram for gram a dozen times more potent than that unleashed by rattlesnakes. It is a complex toxin that affects control of the cardiovascular and muscular system.

You may feel an abrupt stabbing pain when you are bitten. Pain spreads quickly from the bite wound to your midsection, especially around your stomach and back. Cramps and stiffness will follow, along with nausea, increased blood pressure, and labored breathing. Next up is vomiting, irregular heartbeat, convulsions, general irritations and even priapism (unnatural stiffening of the penis . . . if you're a male). Facial muscle spasms, anxiety, pale skin, and an inability to stay still will be followed by a substantial increase in mucous and sputum. Liquid buildup in the lungs, known as noncardiac pulmonary edema, will keep oxygen from getting to your other organs. That means the end of the road, as no oxygen equals no life.

KNOWN BY SCIENCE AS:

The southern black widow (*L. mactans*), the northern black widow (*L. variolus*), and the western black widow (*L. hesperus*), all of which are found in North America. Latrodectism is the medical term for what happens to you after you're bitten.

MEDICAL CAUSE OF DEATH

Hypoxia; possibly cardiac failure

TIME TO KILL

The first signs of trouble occur within twenty to thirty minutes after envenomation. Death can occur after five days if no antivenin is taken.

HIGHEST RISK

People who stick their fingers where they don't belong; landscapers; construction crews in the Southwest; infants; the elderly

LETHALITY

Low, less than 5 percent. Black widows do not usually inject enough venom to kill a healthy adult. Antivenin supplies are prevalent, further minimizing the actual death toll.

KILLS PER ANNUM

Three to six

HISTORIC DEATH TOLL

The only significant study of black-widow-spider deaths, conducted in 1963, showed that sixty-three people died over a ten-year period between 1950 and 1959. Many physicians are unable to detect the cause of death from a black widow.

NOTABLE VICTIMS

In the summer of 2006, a seventeen-year-old boy and a twenty-seven-year-old woman were killed by black widows in the same week. Making it stranger still is they lived in Albania, and the black widows were believed to have come into the country aboard transport ships.

HORROR FACTOR: 1

It is just a bug bite, after all.

GRIM FACTS

- Prior to the widespread use of indoor plumbing, a significant number of black-widow bites were delivered to the genitals and buttocks due to the spiders nesting in outhouses.

- Members of the "widow" family of spiders are found all over the world and are believed to account for the majority of fatal spider bites.

- Throughout the twentieth century, black widow thread was highly prized for use as the crosshairs in scientific instruments and sniper scopes.

BOMBS

Nuclear warheads, artillery shells, missiles, improvised explosive devices, land mines, depth charges, hand grenades . . . it doesn't matter what you call them, they're all the bomb.

HOW IT KILLS

Bombs are weapons designed to kill many people at once, giving them one of the biggest kill ratios of anything that will kill you. They do this by releasing an incredible amount of energy very quickly. This happens by detonating liquid or solid explosive compounds so that they instantly become gaseous.

The gases, reaching temperatures of up to 7,000 degrees Fahrenheit, are formed in less than 1/10,000 of a second after detonation. The blast pressure wave, called a show wave, exerts pressure of nearly 700 tons per square inch on the surrounding atmosphere and moves at more than 12,000 miles per hour.

The explosive container also rips apart, sending shrapnel in all directions. If you are in the immediate blast area, you are likely to die from one of three events. The first is complete annihilation caused by the shock wave ripping you apart. This means removing limbs, head, internal organs, or anything else in less time than it takes to read this line. In that case, you die from simply not being one organism anymore.

The second method of death may also come from the shock wave, but is less grisly and is dependent upon the power of the blast and your distance from it. The wave hits you hard enough that it crushes your heart and lungs, resulting in immediate cardiac arrest and respiratory failure.

Shrapnel is the third cause of death. You may have escaped the incredible pressures of the shock wave, but the flying shards of metal tear through your body, slicing through blood vessels, and cutting open vital organs. You will live for a few moments, but blood loss, shock, and organ failure will take their toll as surely as the shock wave would have.

KNOWN BY SCIENCE AS:
Primary blast injury from a high order explosive

MEDICAL CAUSE OF DEATH
Respiratory failure; cardiac arrest; blood loss; shock; organ failure

TIME TO KILL
Immediate

HIGHEST RISK
Soldiers on foot patrol, political targets; people living in countries torn by civil war

LETHALITY
High. Bombs are designed so that they have a number of ways of making sure that they kill as many people per event as possible.

KILLS PER ANNUM

Thousands, due primarily to military conflicts along with the increasing incidence of suicide bombings in the Middle East since the 1990s.

HISTORIC DEATH TOLL

Millions. War and religious conflict have provided plenty of victims for bombs. Nothing has come close, however, to the more than 100,000 Japanese who were killed by just two atomic bombs in 1945.

NOTABLE VICTIM

Rajiv Gandhi, the former prime minister of India, was killed when a woman detonated a bomb in his face on May 21, 1991. Fourteen others were killed in the attack, one of the first suicide bombings.

HORROR FACTOR: 2

If you're in the direct path of the blast, you never see it coming.

GRIM FACTS

- The largest conventional bomb used by the United States is the MOAB, or Massive Ordnance Air Bomb. It weighs 21,000 pounds and is thirty feet long. MOAB's nickname is Mother of All Bombs.

- Grenades, which are time-delayed bombs delivered by hand, were invented around A.D. 1000 by the Chinese.

- A "dirty bomb" combines a conventional explosive, such as dynamite, with radioactive material. A "smart bomb" uses sensors to find a specific target.

BOTULISM

Botulism used to be something that your mother occasionally warned you about when you were opening six-year-old cans of soup. Today, men and women inject a form of it into their faces to help them look younger. This despite the fact that the American Medical Association calls the botulinum toxin the most poisonous substance known to man.

HOW IT KILLS

The bacteria responsible for botulism is found in the ground all around you. It exists as a harmless spore until it is dug up with vegetables headed for the canning factory and makes its way into cans headed for grocery-store shelves. The spore comes to life in airless conditions—it is anaerobic—and favors vegetables that have a low acid content, such as beets, green beans, spinach, and corn, although preserved fish and meat also provide it a perfect home. Once awakened, it forms a toxin called botulinum, which is the cause of a disease called foodborne botulism. And while eating a bad tin of sardines or rank outdated soup doesn't sound that bad, especially if you're on a budget, you should know that one gram of botulinum has the potential to kill a million people. Two ounces could kill everyone in the United States.

Symptoms begin as soon as six hours after you eat contaminated

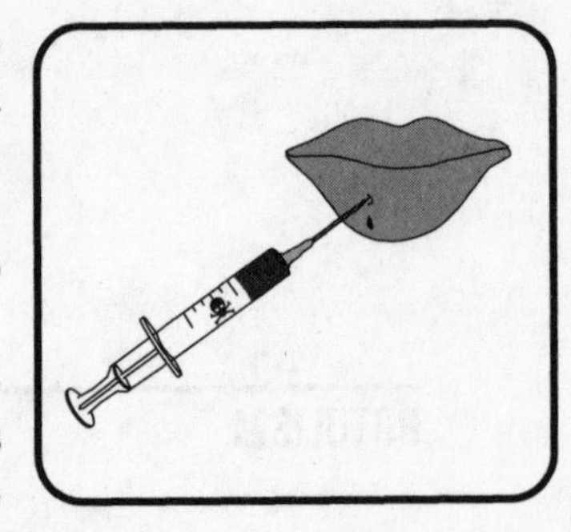

food, but typically show up two to three days later. And then it hits you like a freight train. The toxin attaches itself to nerve endings, effectively blocking the brain's ability to control muscle contractions.

The symptoms are similar to being seriously drunk. Your eyelids droop and you get blurred and double vision. Your mouth goes dry, you start slurring your speech, and you have difficulty swallowing. Vomiting and diarrhea are next up. Then you experience overall weakness in your muscles, which is the onset of full-blown muscle paralysis.

It spreads to your limbs as your arms, calves, and thighs stiffen. Then the muscles in your chest and abdomen tighten, and your respiratory muscles lock up. Breathing becomes difficult, and finally impossible as your throat closes and your lungs cease to suck in oxygen. You do not lose consciousness, and are aware of the paralysis overtaking you. On the bright side, there is no fever.

KNOWN BY SCIENCE AS:

Botulinum is a neurotoxin produced by *Clostridium botulinum*, an anaerobic bacteria. Its root is in the Latin "botulus," meaning sausage, which is the shape of the spores. There are seven forms of botulin, listed A through G; only A, B, E, and F are dangerous to people.

MEDICAL CAUSE OF DEATH

Asphyxiation

TIME TO KILL

Symptoms start as early as six hours after exposure and as late as ten days, but once they start, you've got four days until the coroner shows up.

HIGHEST RISK

Fans of way past the "serve by" date on canned food; people who can their own beets and beans and don't heat the vegetables long enough to kill bacteria; people who get their Botox facials in back-alley clinics.

LETHALITY

Medium—60 percent. If you get to a mechanical ventilator in time, you can greatly improve your chances of survival although you'll be on that respirator for weeks. Antitoxins work well if administered within hours of eating contaminated food.

KILLS PER ANNUM

Estimated at one to three in the United States

HISTORIC DEATH TOLL

Several hundred to a thousand, mostly since the growth of processed foods in the 1950s. Outbreaks tend to come from the same source (canning factory, local restaurant, homemade fixin's at a picnic).

NOTABLE VICTIM

Sam Cochran, a sixty-one-year-old banker, died of botulism in June 1971, after eating canned vichyssoise soup made by the Bon Vivant company. The subsequent recall of more than million cans—of which less than a dozen were said to have been contaminated—bankrupted the company.

HORROR FACTOR: 5

Contracting botulism is hardly the stuff that requires a visceral and immediate reaction, but the gradual paralysis is certain to make you feel like you are slowly being turned to stone.

GRIM FACTS

- Botox is the trade name for the botulinum toxin A as produced by Allergan, Inc. While it is not technically botulism, it contains the exact same toxin, albeit in a diluted form. Botox works much as botulism does, by paralyzing muscles, which helps facial muscles stay firm.

- In addition to foodborne botulism, there are two other forms of botulism: infant botulism, which occurs in the intestines of infants, and wound botulism, in which a wound or sore produces the toxin after getting infected by the *Clostridium botulinum* bacteria.

- Raw honey is a source for infant botulism because it can harbor the spores that produce botulin, so children under one year old should never be given honey.

- Botulinum toxin's potency has led to fears of it becoming weaponized for bioterrorism.

BRAIN ANEURYSM

Pop goes the brain balloon.

HOW IT KILLS

If you've ever blown up a balloon with a weak spot in the latex, you've probably noticed how the thinner area inflates easily and is prone to popping. Aneurysms, in very much the same way, are bulges along a section of blood vessel. And just like a popping balloon, the bursting of an aneurysm can be kind of startling.

Aneurysms in the brain, or cerebral aneurysms, strike fear because they are silent killers. Unless recognized on a brain scan scheduled for some other condition, most people never know they have one until it ruptures. It's estimated that some 15 million people in the United States today are walking around with unruptured brain aneurysms. Walk carefully, people.

Just like forcing too much air into a balloon causes the weak spot to split, increased blood pressure can cause the aneurysm to hemorrhage. Smoking, hypertension, cocaine abuse, and alcohol abuse have all been associated with the rupturing of cerebral aneurysms, though anything that causes a sudden increase in blood pressure—from strenuous activity to a violent sneeze or cough, and maybe even blowing your nose—can cause a burst. The popular conception that you

can be so stressed out that an aneurysm bursts isn't entirely far-fetched; high-strung people often have high blood pressure, and hypertension is a big risk factor.

The first sensation you'll experience is the sudden onset of the most intense headache you've ever had. Vomiting may come next, accompanied by nausea, double vision, a stiff neck, and—probably the most desired symptom at this point—a loss of consciousness. About 15 percent of those struck won't even make it to the hospital alive.

As you might imagine, blood isn't supposed to be spilling all over the place inside your head. When the bubble bursts, brain cells in the vicinity are damaged and destroyed, compromising brain function. The interruption of blood flow can also kill you in the form of a stroke. Furthermore, the spilled blood can lead to hydrocephalus—fluid on the brain—which can increase cranial pressure and push brain tissues up against the inside of your skull, leading to brain damage in the best case and death in the worst. Another late-coming but deadly complication is vasospasm, in which other blood vessels cause brain damage or stroke by contracting and limiting blood flow.

KNOWN BY SCIENCE AS:

Cerebral aneurysm; intracranial aneurysm

MEDICAL CAUSE OF DEATH

Brain hemorrhage; brain trauma; stroke; hydrocephalus; vasospasm

TIME TO KILL

Up to 15 percent die within minutes; 50 percent die within thirty days; remaining patients die within six months after suffering severe

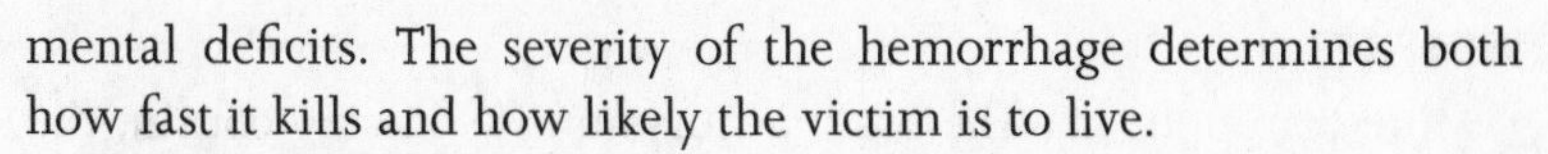

mental deficits. The severity of the hemorrhage determines both how fast it kills and how likely the victim is to live.

HIGHEST RISK

Smokers with hypertension and genetic history of aneurysms

LETHALITY

High. The majority of people with a severe burst aneurysm don't live to tell about it, and among those who do, you can't really understand what they're saying.

KILLS PER ANNUM

10,000 in the United States, by the most conservative estimates

HISTORIC DEATH TOLL

Millions, verging on billions. The number of fatalities by bursting brain balloons have historically tracked closely to statistics on smoking and hypertension.

NOTABLE VICTIM

Actress Phyllis Kirk appeared scared to death of Vincent Price in 1953's *House of Wax*, but it was a cerebral aneurysm that killed her in 2006.

HORROR FACTOR: 7

The out-of-nowhere aspect is shocking and cruel, and to survive frequently means living a life compromised by brain damage.

GRIM FACTS

- More common than the brain aneurysm is its kissing cousin, the aortic aneurysm, which kills about 15,000 Americans every year.
- About 5 percent of the population has an unruptured aneurysm.
- Women are more likely than men to have brain aneurysms by a ratio of 3 to 2.

Final Approach

Wim Delaere was lounging in his parent's house in Belgium on July 4, 1989, when an unmanned Soviet MIG-23 fighter crashed into him. The MIG's pilot had bailed out over Poland during an equipment malfunction and the plane continued all the way across Europe to the Belgian border before running out of fuel. The nineteen-year-old Delaere was killed on impact.

BULLET WOUND

Bang, bang. Shoot, shoot.

HOW IT KILLS

Bullets are small objects made of metal that travel very, very fast. A bullet shot from a high-caliber handgun can briefly reach speeds of over 1,000 miles per hour. That's a lot faster than you can run . . . or zigzag.

Handgun bullets are generally under an inch long, although their size and design vary widely. A bullet wound that does not destroy a vital organ, like your heart or brain, will leave you with a good chance of surviving. But let's say you get hit in the chest, a direct shot. You will not get knocked down, as a bullet isn't strong enough to do that (research has determined that the people who fall after being shot do so because they think that is what's supposed to happen).

The bullet penetrates your skin, ripping open your flesh and tearing capillaries. As it encounters resistance from your flesh and muscle, the tip of the bullet flattens out, expanding its overall width. The shock wave from it entering your body puts extreme pressure on nearby tissue and organs, disrupting their operation.

The bullet now moves forward and shatters your rib cage, spraying shards of bone that pierce organs and blood vessels. The bullet's "crush mechanism" destroys tissue and muscle—leaving a permanent hole—as it penetrates deeper into your body and enters your heart. Blood spills out of your heart, filling up your chest cavity, and gushing liberally from the entry wound in your flesh. Your blood pressure sinks dramatically, and your heart fails from the damage. Coupled with blood loss, you suffer cardiac arrest.

The bullet comes to a stop somewhere in your body—if it has run out of energy—or exits out your back, creating a big hole that will further drain your blood.

KNOWN BY SCIENCE AS:

Deadly force; projectile wounding

MEDICAL CAUSE OF DEATH

Depends on where the bullet hits, but blood loss, brain death, cardiac arrest, and organ failure are all potential candidates

TIME TO KILL

Several minutes, if you're lucky. If the bullet damages a vital organ but doesn't kill you right away, it could be weeks or months until you die.

HIGHEST RISK

Gang members; drug dealers; soldiers; hunters; people living in "the bad part of town"

LETHALITY

Low. Unless the bullet enters your brain, heart, or lungs, the chances for surviving are very high.

KILLS PER ANNUM

Approximately 100,000 worldwide; about 8,000 annually in the United States

HISTORIC DEATH TOLL

Millions. Homicide, suicide, war, and accidents have ensured guns and their bullets an exalted status in the history of killing.

NOTABLE VICTIM

President John F. Kennedy was killed on November 22, 1963, by one, two, or three bullets—depending on whom you believe—fired by one gunman or more from a distance of 265 feet—again, depending on whom you believe. Either way, the bullets tore through his throat, sheared away much of his skull, and destroyed his brain. The behavior of these bullets has been scrutinized by more people than you can count.

HORROR FACTOR: 9

"Oh my God. I've been shot!" is one of those phrases you really hope you'll never have to utter in your lifetime.

GRIM FACTS

- The FBI actually produced a report stating that short of a direct hit on the brain, the spinal cord, or the heart, a single bullet from a handgun is not enough to fully incapacitate a criminal.

- Handguns are notoriously inaccurate at ranges beyond fifty feet.

- Hollow-point bullets are designed to expand inside the human body, tearing through as much tissue as possible. They were outlawed for use in war by the St. Petersburg Declaration of 1868 and the Hague Convention of 1899.

- Small-caliber handguns and semiautomatic weapons are designed specifically to kill human beings and are useless for sport hunting because they are either ineffective or do too much damage to the animal.

BURNING AT THE STAKE

Fire is at once both a great purifier and a great destroyer. Most of all, however, it is a great burner.

HOW IT KILLS

Almost anything, including steel, will melt if subjected to extremely high heat. Usually this heat comes from wood, paper, gasoline, oil, and a host of other flammable materials. In the case of a public burning, body fat provides a highly unusual fuel for the flames.

Let's say your neighbors have decided that your personal success in life is due to the fact that you've either sold your soul to the devil or you are a witch. They've also decided that they're not too happy about having you live in their subdivision. Instead of talking this out rationally with you, they choose to remedy the situation via the quickest and most efficient route: they're going to burn you at the stake. Right now.

Dragging you kicking and screaming to the town square, they tie you to a tall pole. Because it is hard to get the human body to ignite, they pile wood up around your feet and light that. If the resultant fire is big enough and hot enough, you'll suffocate from all the oxygen that the fire is sucking into itself. If it's a slow burning fire, you are in for a long trip through a personal Hell on Earth.

As the temperature of the fire climbs over 300 degrees Fahrenheit, the flesh on your feet and legs will start to blister. The flames that lick higher, up around your thighs and hands, will start toasting your skin, causing it to char and blacken. Your hair will singe and crinkle. Since your body is primarily made of water, the heat will cause the cells in your body to start steaming away liquids. Your clothes will catch on fire after your skin has already started smoking.

As the temperature increases and the moisture in your skin dries up, your flesh itself will start burning. You will go rapidly from first-degree burn (like a bad sunburn) to second-degree (deep burn into the flesh) to third-degree (full-thickness destruction of the flesh) all the way to fourth-degree (burning past the flesh and into the underlying tissue). At 480 degrees Fahrenheit, the fat in your body will become like candle wax and fuel the fire without the need for wood.

Along the way, your capillaries will burn up, preventing blood from getting to your skin. But that won't matter, because blood will spill out of your body due to the fact that no flesh is holding it in. Your nerves—after signaling indescribable pain to your brain—will fry along with your charred skin. Blood loss will lead to low blood pressure, and if you're lucky your heart will give out. If you're not, then other organs will burn, body fluids will vaporize, and you'll hope against hope that this scenario is not the one that's awaiting you when you enter the afterlife—which will be just moments away.

KNOWN BY SCIENCE AS:

Thermal burn

MEDICAL CAUSE OF DEATH

Shock; blood loss; organ failure

TIME TO KILL

Several minutes to several hours, depending on the temperature and concentration of heat

HIGHEST RISK

Witches; heretics; saints; sinners; political dissidents and traitors

LETHALITY

High—100 percent. Once the flames are rising, the phrase "time to burn" no longer applies to you.

KILLS PER ANNUM 

Ten to twenty. While people are no longer burned at actual stakes, per se, they are executed by similar methods such as "necklacing" (placing a gasoline-soaked tire around the neck of the victim), and locking them in their homes and torching the structures.

HISTORIC DEATH TOLL

Thousands upon thousands. Killing with fire has been used since the dawn of history by almost every society from the primitive Aztecs and Romans on to the allegedly civilized French and British.

NOTABLE VICTIM

After trying to save France from "godless" British oppressors, nineteen-year-old Joan of Arc was convicted of being a heretic. The

Brits then burned her at the stake on May 30, 1431, making teenage Joan the definitive poster child for flaming out.

HORROR FACTOR: 10

The highest. It's bad enough when you burn your fingers on a match; imagine that match is a giant blowtorch and it's aimed at your entire body.

GRIM FACTS

- The human body must be heated to over 1,400 degrees Fahrenheit for all tissue, organs, and bones to be cremated.
- More than 5,000 people die each year in the United States from a burn-related injury.
- Spontaneous human combustion (SHC), a condition in which the body ignites itself without external combustion, has been widely reported and investigated. There is no proof that this phenomenon is, in fact, real.

CANCER

Incidence is dropping and survivability is on the rise. But a cancer finding remains the most dreaded diagnosis of a visit to the doctor.

HOW IT KILLS

Cancer. Few words in the English language are as black and foreboding. You can't even get away with a good cancer joke outside of an oncology ward, so let's just get through the facts and turn the page.

The smallest building block of the body is the cell, and different types of cells make up our entire anatomy. In a healthy person, cells grow, divide, and die in a natural and orderly process. But sometimes the controls and mechanisms of cellular life malfunction, and cells form when they shouldn't or don't die when they should. Rogue cells form masses of tissue, or tumors, which can be either benign (noncancerous, and often harmless) or malignant (cancerous). Malignant tumors invade and destroy the healthy tissues around them.

Concern mounts when malignant cell groups reach the lymph nodes, because the lymphatic system contains a pathway of vessels that reaches all of the body's organs. Similarly, once cancer is in the bloodstream, all of the body's tissues are vulnerable to new tumors.

The spread of cancer is known as metastasis. Any tissue within any organ of the body can become cancerous.

Cancer isn't just one disease, but a whole damned family of related conditions. It breaks down into three main categories: *Carcinomas,* the most prevalent group, are cancers of cells that line or cover organs. They include cancers of the lung, colon, breast, prostate, stomach, and thyroid. *Leukemias* and *lymphomas* are cancers of the blood. *Sarcomas,* which occur more frequently among the young, are cancers in muscle and connective tissue, such as bone cancer.

How cancer kills you depends on the tissues it destroys and the vital systems it disrupts. The possibilities are as vast and complex as our physiology. In the digestive system, it can prevent absorption of food and starve you of nutrients. In the brain, it can cause stroke or the seizure of autonomous functions. In the lungs, it can prevent sufficient oxygen from being absorbed. In the liver or bones, it can promote excess calcium production (hypercalcemia); calcium then builds up in the blood and can lead to kidney failure, heart arrhythmia, or dysfunction of the nervous system. Cancer has a million tricks.

In nonvital organs such as the breasts, eyes, or testicles, it is not local malignancy but the potential for metastasis that is life-threatening. When other options fail, oncologists would sooner have you survive with one less body part than risk the spread of cancer to your other vital organs.

KNOWN BY SCIENCE AS:

Cancer

MEDICAL CAUSE OF DEATH

Organ death; dysfunction leading to failure of a vital system

TIME TO KILL

Varies greatly according to type and at what stage a patient has been diagnosed. More than a third die within five years of diagnosis.

HIGHEST RISK

Family history and age are uncontrollable factors, but certain lifestyle habits are known to increase risk. Smoking, drinking too much, having unsafe sex, excessive sun exposure—that is to say, all the fun stuff—can give you cancer.

LETHALITY

Very high, but dropping. Every year more cancers are treatable or survivable. According to statistics quoted by the National Cancer Institute, the five-year survival rate is currently 64 percent.

KILLS PER ANNUM

Estimated at over 550,000 in the United States, approximately 8 million globally.

HISTORIC DEATH TOLL

Countless millions; humans have been succumbing to carcinogens since the earliest mammoth barbecues. But there's a silver lining: cancer incidence on the whole has been declining since the early 1990s, though several types are on the rise (including liver, esophagus, lukemia, brain, bladder, and more).

NOTABLE VICTIM

John Cheever died in 1982, at the age of seventy, after a month-long battle with cancer that had spread from one of his kidneys. While undergoing treatment, the author was quoted as saying, "My veins are filled once a week with a Neapolitan carpet cleaner distilled from the Adriatic and I am as bald as an egg. However, I still get around and am mean to cats." Leave it to Cheever.

HORROR FACTOR: 8

Having one's own body turn on itself is internal terrorism.

GRIM FACTS

- Cancer is the nation's Number 2 killer, second only to heart disease.

- More than half the people diagnosed with cancer today will not die of the disease.

- The rate of pesticide levels in human blood is on the rise.

- Without encroaching on the territory of Patch Adams here, let's try one cancer joke and see how it goes. The doctor says, "I have your test results here and I'm afraid there's very bad news. Not only do you have cancer, you also have Alzheimer's." The patient says, "Phew—I was afraid I had cancer!"

CAR VERSUS TREE

Buckle up, don't drink up.

HOW IT KILLS

To take the long, dark view on death in America, disease wipes out the elderly population while car crashes take care of the young. The National Highway Traffic Safety Administration reports that motor-vehicle crashes are the leading cause of death for nearly every age group from three to thirty-four, holding sway until cancer claims the throne for those aged thirty-five to sixty-four.

We can't blame it all on other drivers, either, as the number of fatalities in single-vehicle accidents is not far outshone by multivehicle crashes. In 2006 alone, 12,327 fatalities involved a single car driving off the roadway, just a few thousand shy of the number killed in multiple-vehicle accidents occurring on the road. A third of those killed struck a fixed object such as a tree, a lamppost, or a bridge. Those who spent their last

moments chatting on a handheld phone now know that the Reaper never screens his calls. He'll pick up every time.

Nor can we fault bad weather or the traffic on superhighways, since most fatal accidents occur in clear weather on the nation's two-laners. Rather, it's having a belt of alcohol—but not a belt across your shoulder—that takes the highest toll. Deaths from alcohol-related crashes hover around 17,600 every year. In a strikingly similar statistic, about 16,000 die unrestrained by a belt. Seat belts or other restraints go unused in 55 percent of all fatal crashes.

Car-crash deaths—like airplane, train, and motorcycle deaths, as well as deaths by falls and beatings—are most often linked to fatal impact injuries. We have learned from crash-test dummies that less damage is done at lower speeds, or when an airbag slows down a body as it encounters a stationary object. Both methods of gradual deceleration increase the chance that your head will deflect off an object without too much harm. But in the case of rapid deceleration, your head moves from speed to complete stop in a shorter span of time. On a time-lapse camera, the back of a dummy's head can be seen still moving forward after the front of it has met the wall. It's really not fun for the dummy.

You're no dummy made of plastic and pads, so you also have to consider how much compression your body can sustain in a tête-à-tête with an oak tree. To understand how compression affects the body, imagine bouncing a bowling ball on a trampoline; the rubber floor can compress several feet, deflect the ball, and return to shape with no damage. Now imagine that your chest is the trampoline, and the bowling ball is an engine block headed through the car's firewall and into your chest. The muscles and bones in your torso can typically compress about two inches under the engine without causing fatal damage. Much beyond that, they don't bounce back.

KNOWN BY SCIENCE AS:

Fatal single-vehicle collision with a fixed object

MEDICAL CAUSE OF DEATH

Blunt force trauma; fatal crush injury; hemmorhage; decapitation

TIME TO KILL

Instantaneous on impact, with variation for untreatable fatal injury

HIGHEST RISK

Unrestrained drivers and passengers traveling late at night. More than 70 percent of fatalities involving occupants without seatbelts occur between midnight and 3 A.M., and more often than not on a weekend.

LETHALITY

Medium. Rises in association with speed, seat-belt use, and intoxication.

KILLS PER ANNUM

More than 12,500 fatal single-vehicle collisions with a fixed object

HISTORIC DEATH TOLL

America is getting better at staying alive behind the wheel. Since 1975, the vehicle occupant fatality rate has declined by more than 25 percent.

NOTABLE VICTIM

Artist Jackson Pollock hadn't created any of his famous splatter art in the year 1955 until August 11, when he drunkenly rammed his Oldsmobile convertible into a tree in Springs, New York.

HORROR FACTOR: 5

Barring the awareness that your soft flesh is about to meet metal, glass, or bark with life-ending force, suffering should be minimal provided you don't survive too long after impact.

GRIM FACTS

- The nation's capital has the highest rate of alcohol-related accidents. Fifty-eight percent of all fatal crashes in Washington, D.C., involve alcohol.

- More than 12 percent of drivers admit to having driven under the influence of a substance at least once in a year's time. Among people aged 21 to 25, more than 27 percent admit the same thing.

COPD

If lung cancer isn't enough of a deterrent, the risk of COPD may keep you from reaching for another smoke.

HOW IT KILLS

How can the nation's Number 4 killer, chronic obstructive pulmonary disease, have slipped under the radar? If you haven't heard of COPD, you may know of its more familiar subcategories: chronic bronchitis and emphysema. Both are potentially deadly and both are typically brought on by smoking. Plus, COPD is one of those two-for-one type diseases—most people with COPD have both bronchitis and emphysema.

While the rates of other leading killers like heart disease and stroke are declining, COPD is on the rise. By the year 2020, COPD is expected to take over the Number 3 slot. Given the nation's love affair with cigarettes, it is unlikely the trend will reverse.

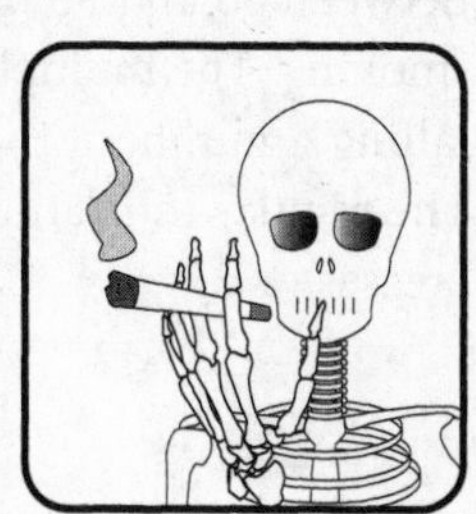

Let's say that you've smoked a pack a day for the past twenty-five years. The damage done by COPD is already present in the smallest, deepest reaches of your respiratory system. Normally, oxygen and carbon dioxide are exchanged between your lungs and bloodstream at the ends of airways, where tiny sacs called alveoli fill with air when you inhale and deflate when you exhale. In a case of chronic bronchitis, the walls of the small airways, which can be as narrow as a hair, become inflamed and swollen shut. With emphysema, the delicate alveoli are damaged so badly, they cannot inflate or deflate at all. The consequent inability to exhale deeply enough leaves air trapped in your lungs, so you can't get enough good air in or bad air out. Starved of oxygen and overwhelmed with carbon dioxide, your respiratory system eventually fails.

KNOWN BY SCIENCE AS:

Chronic obstructive pulmonary disease

MEDICAL CAUSE OF DEATH

Respiratory failure

TIME TO KILL

Studies suggest that approximately 25 percent of people with advanced COPD die within five years of diagnosis. Mortality rates increase with age and lifetime history of smoking.

HIGHEST RISK

Between 85 and 90 percent of all COPD cases are caused by cigarette smoking. The multiple of packs-per-day times years of smoking, totalling a number of "pack years," determines level of susceptibility. The regular inhalation of pollutants, as at some workplaces, also increases risk.

LETHALITY

High. COPD claims the life of a U.S. citizen once every four minutes.

KILLS PER ANNUM

More than 120,000 in the United States, upwards of 3 million around the world.

HISTORIC DEATH TOLL

Half a billion is a conservative estimate. The number of deaths by COPD has doubled in the past twenty-five years, and the WHO estimates that there could be a billion more by the end of the twenty-first century.

NOTABLE VICTIM

Johnny Carson once quipped, "I know a man who gave up smoking, drinking, sex, and rich food. He was healthy right up to the day he killed himself." Maybe Carnac the Magnificent saw it coming. Carson, a longtime smoker, died from emphysema complications in 2005.

HORROR FACTOR: 4

Living with severe COPD is a breath-to-breath struggle for fresh air.

GRIM FACTS

- COPD is way deadlier than the LAPD or the NYPD.
- Rates of COPD are rising faster among women than men, most likely due to the increase in female smoking.
- Current estimates are that 24 million Americans have COPD. Half of them don't know it.

The Magic Bullet

On March 31, 1993, actor Brandon Lee was to be shot by a "fake" bullet while filming the movie *The Crow*. However, another cartridge had accidentally lodged in the gun prior to the scene, and it was that bullet that was fired into Lee's stomach. It hit his spine, and Lee, the son of martial-arts legend Bruce Lee, died several hours later.

DEATH CAP MUSHROOM

Mushrooms are enjoyed as food all over the world, and have found no small amount of fame as a hallucinogenic drug. But eating one particular kind of mushroom will kill you unless you can find an organ donor with a healthy liver really quickly. The name is a dead giveaway.

HOW IT KILLS

There are some fifty species of poisonous mushrooms, but they all pale in comparison to the death cap. The death cap mushroom is found in Europe, the Americas, and Australia. It has an exaggerated toadstool shape (toadstool being the general term for poisonous mushrooms), with a cap measuring more than two inches across and a stalk of about six inches. It resembles the straw mushroom, a delicious bit of fungi that you will likely have mistaken the death cap for when you ingest it.

Eating a single mushroom can be fatal because each one contains a significant amount of a toxin called amanitin, which destroys cells by preventing the creation of proteins.

The cells most susceptible to this toxin are in your liver, which is the first organ in the line of fire after you ingest the death cap, followed shortly thereafter by cells in the kidneys and then your brain.

As your liver and kidney cells start dying, the initial symptoms of poisoning start manifesting themselves. These include cramps, vomiting, and watery diarrhea. You will experience extreme thirst and stop urinating. You get a temporary reprieve for a few days until the next wave hits. And it's a tsunami. At this point, your liver starts melting as dead cells in the organ dissolve. Your skin turns yellow as jaundice spreads throughout your body. You may have seizures and delirium, a distinct inability to think straight.

The substances that are normally cleaned out of your body by your liver will start to back up like a clogged toilet. This leads to hepatic encephalopathy, which is brain damage caused when your blood carries toxins, which are usually filtered out by the liver, all the way to the brain. You will go into a coma, one if not several of your organs will fail, and you will die.

One of the only ways to survive the organ damage caused by death cap poisoning is to get a liver transplant. Ditto for your kidneys. Hopefully you have a big family or really good friends.

KNOWN BY SCIENCE AS:

Amanita phalloides is the scientific name for the death cap; the mushroom produces an amatoxin known as amanitin.

MEDICAL CAUSE OF DEATH

Renal failure; cardiac arrest; brain hemorrhage

TIME TO KILL

One to two weeks. The toxins in the death cap shut down your liver and then kidneys, so it all depends on how well you can manage day to day without each one.

HIGHEST RISK

Hippies who aren't picky about where they get their 'shrooms; amateur mushroom pickers; diners who are served by dim-witted chefs

LETHALITY

Medium. Roughly 20 percent if diagnosed within a day, rising to over 60 percent if not caught within three days.

KILLS PER ANNUM

One or two in the United States, estimated at up to several dozen worldwide

HISTORIC DEATH TOLL

Not known, but probably well up into the hundreds of thousands, as deaths have occurred every year since the first intrepid gourmand ventured to taste the fungi.

NOTABLE VICTIMS

In 1740, Holy Roman emperor and king of Hungary Charles VI died after eating sautéed mushrooms widely believed to be death caps. His demise led to the War of Austrian Succession, better known as "King George's War."

More recently, on January 11, 2007, eighty-three-year-old Miranda Epifania died from eating a death cap after taking her family mushroom picking in Santa Cruz. The other five members of her family survived.

HORROR FACTOR: 4

Not too bad until you start having seizures from organ failure

GRIM FACTS

- Commenting on the war that followed the death of Charles VI, Voltaire wrote that a simple dish of mushrooms changed the destiny of Europe.

- Many victims of the death cap in the United States are immigrants from Laos, Cambodia, Thailand, and Vietnam, who mistake the mushroom for edible favorites found in their homeland.

- An increase in death cap poisonings in California has been blamed on would-be gourmets looking for unique foods to add to their home menus.

- Blood transfusions, dialysis, and liver and kidney transplants may all be required to save the life of a death-cap victim.

DEHYDRATION IN THE SAHARA DESERT

The combination of high temperatures and low water supply does a double whammy on our H_2O-based physiology. When you can't find a drink in the Sahara, the world's largest desert leaves you fried and dried.

HOW IT KILLS

Without much effort at all, adults lose about ten cups of fluid every day in the course of breathing, sweating, and having a whizz or two. Unless this small percentage of the body's fluid (1 to 2 percent) is replenished by drinking water, the earliest symptoms of mild dehydration—thirst, dry mouth, decrease and/or darkened urine output—will appear after exercise or simple lazing about on a hot summer day.

To gain the slightest sense of what a more serious case of dehydration feels like, consider your worst hangover ever. You wake with a desperate thirst, and because your body has been sponging up the fluid available in saliva and mucus, you feel like you've slept with a dirty ski sock in your mouth. Though your mind is functioning well (at least better than it was the night before), your brain is unable to calibrate a normal sense of balance, so you feel light-headed and dizzy. Your muscles are weak. If they're strong enough to carry you

to the bathroom, you will notice that your urine is almost exactly the hue of a blended Scotch. But trumping all of these symptoms is a dry, crushing headache that feels as if you've been shot through the temples. You're unable to stay upright for long.

Ethanol's capacity to dehydrate the body is partially to blame for your hangover, but a night on the town is nothing compared to a waterless day in the desert. Temperatures can soar well over 110 degrees Fahrenheit on an afternoon in the Sahara. Caught between the baking sand and the arid air, your body begins drying out like a cake left too long in the oven. At first you sweat, but the moisture evaporates before it can cool you down. Without water intake, soon you can't produce any perspiration at all. You have no spit, no urine. When you pinch yourself to see if you're still alive, the gathered skin no longer snaps back into place. You want to cry, but you have no tears. The desert is sucking you dry.

More than 15 percent of your body weight can be lost on the way to death, making dehydration the world's most effective but least desirable weight-loss plan. Even your brain starts to shrink within your head. Normal electrical exchanges in your brain become erratic, leading to seizures or sending you straight into a coma. Since the reduction of bodily fluids includes your blood, pressure within the circulatory system drops and reduces the amount of life-giving oxygen that can reach the body's organs and tissues. Your kidneys, unable to filter the blood and rid the body of excess fluids and waste, begin to shut down. Any of these systemic failures may spell your end—which is bad news for you, but good news for the patient buzzards who've been circling since you left Tunisia.

KNOWN BY SCIENCE AS:

Hypohydration

MEDICAL CAUSE OF DEATH

Brain damage; hypovolemic shock; kidney failure

TIME TO KILL

Ten to forty-eight hours

HIGHEST RISK

Children and the elderly in families attempting a desert crossing without supplies

LETHALITY

High. No one can survive the desert without adequate water intake.

KILLS PER ANNUM

Hundreds, especially in years when citizens flee North African countries by force or in desperation

HISTORIC DEATH TOLL

Undetermined, but likely in the millions. Throughout history, scores of exiles, refugees, and immigrants have attempted Sahara crossings to reach the Red Sea or the Mediterranean.

NOTABLE VICTIMS

In 2003, a newspaper in Ghana reported that more than 200 people had died from dehydration or "fatigue" in the Sahara. The Ghanaians were trying to cross on an ill-advised route to Europe through Libya.

HORROR FACTOR: 6

Slow and painful torture, courtesy of Mother Nature

GRIM FACTS

- Some victims survive the desert, gulp down several drinks, and then die of water intoxication. Since you lose not only fluids but sodium when we're you're dehydrated, it takes your body a while to reestablish the water/sodium balance in your blood cells. In the condition known as hyponatremia, the rapid consumption of liquids causes too much water to be absorbed by the cells, which can lead to a fatal swelling of the brain.

- Most dehydration fatalities are caused not by desert crossings but by acute diarrhea.

DOGS

Dogs are man's best friend—except when they're not. Being carnivores, dogs are equipped with sharp teeth and strong jaws. That's the pretty side of the "bad dog" equation, because dogs also carry rabies.

HOW IT KILLS

All dogs will bite, some more savagely than others. Few breeds will actually mount an attack that results in the death of a human, as the relationship between the two species has evolved to one of mutual tolerance. Most of the dogs that do assault people are those that have been bred or trained to attack or fight in competition.

Like other predators, dogs kill using their teeth and claws. On prey the size of a person, they will lunge for the throat and attempt to sever the windpipe and major arteries, resulting in asphyxiation and blood loss. As with other mammalian predators (*see* **Lions, Tigers, and Bears**), the mauling involves lots of torn flesh and copious amounts of spilled blood.

That said, what makes dogs a

real danger to humans is their ability to spread rabies. The rabies virus can be transmitted from dogs to people through bites because the virus lives in the saliva of an infected dog. Dogs typically get infected by other animals that have scratched or bitten them. A rabid dog will show signs of heightened aggressiveness through snarling and bared teeth, as well as foaming saliva.

Once Old Yeller has bitten you, the rabies virus works its way into your central nervous system. Initial symptoms are somewhat flulike, with fatigue, headache, and fever all showing up and lasting for a few days. During this time, the virus is attacking the protective sheaths covering the nerve cells in your brain and spinal cord, where it steps up its growth. You'll start to feel disoriented, start having hallucinations, and find it hard to fall asleep.

Lack of muscle control leads to paralysis and an inability to breathe and difficulty in swallowing. The combination of not being able to swallow and deteriorating mental condition will cause hydrophobia, an extreme fear of water, and one of the strangest symptoms in this book. You then slip into a coma, cursing the day you brought that mangy mutt home from the pound without making sure it had all its shots.

KNOWN BY SCIENCE AS:

Domesticated dogs are *Canis lupus familiaris*. Death by dog is referred to in the medical community as dog-bite-related fatalities (DBRF). Rabies is called hydrophobia due to one of its symptoms, and is caused by the lyssavirus.

MEDICAL CAUSE OF DEATH

Respiratory failure; brain disease

TIME TO KILL

A dog mauling will kill you in less than ten minutes. Rabies will kill you a week to ten days after symptoms show up, which can be about two weeks after the actual bite occurred.

HIGHEST RISK

Backwoods dog breeders; gangsta rappers with ill-tempered pets; street urchins who want to pet the nice doggy with the sudsy mouth

LETHALITY

As high as it gets, if the dog is rabid. Once the symptoms show up, you're already dead. Unless you received a vaccine prior to the bite, or had the bite cleaned and scrubbed to prevent bacterial infection, rabies is a death sentence.

KILLS PER ANNUM

From 40,000 to 70,000 rabies deaths worldwide, primarily in Asia, Africa, and Latin America, all of which have large populations of stray dogs.

HISTORIC DEATH TOLL

Millions. As long as man has been domesticating dogs, he's been dying from rabies. And occasionally getting chewed up by his spike-collared pit bulls.

NOTABLE VICTIMS

On January 26, 2001, thirty-three-year-old Diane Whipple was attacked by a neighbor's Presa Canario dogs in the hallway of her San Francisco apartment building. The two huge dogs—weighing over one hundred pounds each—ripped Whipple apart, without provocation, which resulted in her death and a conviction of involuntary manslaughter and second-degree murder for the dogs' owners.

Edgar Allan Poe, long thought to have died in 1849 from complications of alcoholism, may have actually died from rabies, according to a 1996 report in *The Maryland Medical Journal.*

HORROR FACTOR: 9

Something about having Fido turn on you while his mouth is foaming and his fangs are bared goes against everything we like about out favorite pets. The response is visceral.

GRIM FACTS

- The Centers for Disease Control (CDC) found that "pit bull–type" dogs accounted for as many as a third of all U.S. dog-bite-related fatalities, more than any other breed.

- Dogs are a significant "disease vector" in third world countries, just as mosquitoes are.

- Rabies can occur in any mammal. While dogs account for most of the third-world transmission, in North America and Europe rabies is most likely to come from bats, raccoons, skunks, foxes, the occasional dog, and even cats.

- The CDC reports that there have been only six documented cases of human survival from clinical rabies.

DIABETES TYPE 2

Nearly two-thirds of the United States is overweight, a primary risk factor for Type 2 diabetes. Some people are biologically predisposed to this incurable disease, while others simply need to say "no" when given the option to super-size another greasy meal.

HOW IT KILLS

One consequence of America's twenty-year slide toward an obesity epidemic has been a swell in the number of people with Type 2 diabetes. Another is the increase in orders for plus-sized coffins. For anyone wondering if the two are connected, the answer is a big fat yes.

Diabetes is characterized by the inability of the body to manage glucose, or sugar. Your body needs glucose for energy, and with the help of the hormone insulin, glucose gets converted into fuel for your cells. But if you have Type 2 diabetes, the type accounting for 90 to 95 percent of all diabetes cases, your body fails to use insulin properly or produce it in enough quantity. That

leaves a surplus of sugar in your bloodstream. As a result, your blood becomes something like Aunt Jemima's maple syrup, which is swell on pancakes but really wreaks havoc on your system.

Because glucose supplies energy to the brain, seizures and coma can result from high or low swings in blood sugar (hyperglycemia and hypoglycemia, respectively). But if you're like 65 percent of the other people with Type 2 diabetes, you'll be killed by heart disease or stroke. Although Type 2 diabetes is the nation's Number 6 killer all on its own, it weighs in heavily by contributing to other leading causes of death, too.

KNOWN BY SCIENCE AS:
Noninsulin-dependent diabetes

MEDICAL CAUSE OF DEATH
Angina; atherosclerosis; stroke; seizure; kidney failure; secondary causes

---

TIME TO KILL
As long as several years

HIGHEST RISK
Those who are over age forty-five and overweight or obese, especially people with family history of diabetes. Individual races, notably blacks, Hispanics, American Indians, and Asian Americans appear to be most susceptible.

LETHALITY ☠☠☠
High. At almost every age, risk of death is twice as great for people with diabetes as for those without.

KILLS PER ANNUM

More than 250,000 in the United States

HISTORIC DEATH TOLL

Type 2 diabetes contributes to more deaths than it directly causes, so numbers are hard to come by. Even though more people die from Type 2 today than they did a century ago, it's safe to say that over 100 million have succumbed to the worst possible kind of "sugar shock."

NOTABLE VICTIMS

Michael "Time to make the doughnuts" Vale, the actor known for his role as a sleepy baker in the Dunkin' Donuts commercials, died in 2005 of diabetes. Type 2 diabetes also sent Syd Barrett, cofounder of Pink Floyd, to the dark side of the moon.

HORROR FACTOR: 3 

The day-to-day concerns of keeping diabetes at bay are eclipsed only by the prospect of your family needing an entire football team to be your pall bearers.

GRIM FACTS

- In 2005, Louisiana was the state with the highest death rate attributed directly to Type 2 diabetes, with more than 1,700 deaths. Maybe it was the fried hush puppies. In Alaska, where almost everyone eats fish, there were only 93 diabetes deaths that year.

- According to the American Diabetes Association, 7 percent of the U.S. population has diabetes—14.6 million have been diagnosed while another 6.2 million people are unaware they have the disease. An additional 170 million are affected worldwide.

- African Americans are twice as likely to die from Type 2 diabetes as white Americans.

It's Got a Kick to It

Jack Daniel, the man who gave the world Jack Daniel's Tennessee whiskey, kicked his company safe when he couldn't remember the combination to open it. He broke his toe, which became infected. Daniel then contracted gangrene and his leg had to be removed, although by that time his blood was infected. He died on October 9, 1911.

DRINKING BINGE

When intoxication turns truly toxic, that's alcohol poisoning. And that can kill you. On the plus side, you won't wake up with a hangover.

HOW IT KILLS

Drinking alcohol is for some an occasional indulgence and for others a regular part of leisure. For others yet, it's a sport. Hard partiers and binge drinkers are prone to drinking to such excess that blackouts and painful morning-afters become a usual part of life. Provided they don't get in a car or do something life-threateningly stupid, most students live to tell the tale (or at least the parts they remember). But many regularly endanger themselves by testing the body's limits. When the body fails the test and a night of drinking tips from inebriation to poisoning, it's game over.

Most people enter the danger zone when blood alcohol concentration (BAC) reaches around .2, which means there are .2 grams of alcohol for every deciliter of blood (compare that to .08, the legal BAC limit for driving in most states). Group drinkers "keeping up" with one another don't typically consider that BAC is impacted by body mass and gender. This is why a 105-pound cheerleader gets drunk

much faster—and more rapidly increases her BAC—when drinking the same amount of alcohol as a beefy jock. A 180-pound male could reach a lethal BAC of .32 after drinking a dozen shots in one hour.

The problem is that the ethanol in alcohol not only lowers your inhibitions and makes you dance funny, it also inhibits routine brain activity. Involuntary breathing functions are affected by high concentrations of ethanol in the bloodstream, resulting in a condition known as respiratory depression. Breathing first becomes shallow and slows drastically, and then carbon dioxide accumulates in the bloodstream. Excess carbon dioxide eventually causes your body to shut down, coma-style. Then you stop breathing entirely.

Another involuntary function compromised by excess alcohol is the gag reflex. When you hear stories about people choking to death on their vomit, it's not because they were too inebriated to turn over and puke on the floor, it's because the gag reflex—the natural response to cough out or repress vomit—is shut down.

KNOWN BY SCIENCE AS:

Ethanol poisoning

MEDICAL CAUSE OF DEATH

Asphyxiation

TIME TO KILL

One to two hours after fatal BAC level is reached

HIGHEST RISK

Binge drinkers with low body mass; users combining drugs and alcohol

LETHALITY

Medium. Many survive alcohol poisoning, though they are subject to brain damage if breathing has stopped.

KILLS PER ANNUM

More than 300 in the United States compared to 40,000 in Russia

HISTORIC DEATH TOLL

Several million. Even though alcoholic beverages have been around for about 10,000 years, most humans have the good sense to stop drinking before dying.

NOTABLE VICTIM

Led Zeppelin drummer John Bonham lost control of his gag reflex after a day of hard drinking. "Bonzo" had started his morning with the equivalent of sixteen shots of vodka at breakfast. And here we thought too much coffee was trouble.

HORROR FACTOR: 1

Most victims will never be the wiser.

GRIM FACTS

- A binge drinker may remain upright and conversant after ingesting a fatal amount of alcohol. While the sensation of being drunk is catching up, the drinker's blood and body tissue are being soaked with the deadly dose.

- There is no way to shortcut the process of metabolizing alcohol; only time can bring blood concentration back to normal levels. Hangover remedies may subdue a headache but they can't expedite the exit of ethanol from the bloodstream.

- When someone poisoned by excess alcohol arrives in the emergency room, all doctors can do is help them breathe, as by intubating, and hope the overdose of ethanol isn't causing permanent brain damage.

A DRINK OF DRAIN CLEANER

Drain cleaners can clear metal pipes clogged with tar-like residue and leave them with a sparkly shine. Imagine what they can do to the fleshy pipes of your intestinal tract.

HOW IT KILLS

If you're assuming it's acid that chews away at stubborn clogs, you may be forgetting about the other end of the pH scale. Drano and its kin are actually strong alkalines, not acids, with a pH hovering around 14. To put that into perspective, consider that water is a neutral 7 (halfway between acid and base), and ammonia is about an 11. Each consecutive number on the pH scale represents a tenfold difference in concentration. If you were to leave a teaspoon of drain cleaner in your hand for about two seconds, you'd feel the early stages of a chemical burn and be left with an irritating rash. Please don't do this.

The active ingredient in most drain cleaners is sodium hydroxide, also known as lye. The chemical produces substantial heat when it comes in contact with moisture, whether that's in your drain or in your gullet.

After drinking drain cleaner, you'll first experience an intense corrosive burning in your mouth, tongue, esophagus, and stomach. You'll start spontaneously vomiting in an effort to clear the poison out of your body, but the cleaner will do just as much damage on its return trip as it did going down. Your throat will swell and breathing will become difficult. Once the drain cleaner reaches your intestinal tract, it will perforate your intestinal lining, and cause hemorrhaging and a narrowing of your gastrointestinal passages. In case reports, causes of death are cited as shock, infection of corroded tissues, lung damage, or simply "loss of measurable pulse."

KNOWN BY SCIENCE AS:

Acute sodium hydroxide poisoning

MEDICAL CAUSE OF DEATH

Shock; irreversible tissue corrosion

HIGHEST RISK

Children under five with access to improperly stored chemicals

LETHALITY

High, primarily among children

KILLS PER ANNUM

Thirty-three accidental pediatric deaths by household poison in the United States (of all types, not exclusively drain cleaners)

HISTORIC DEATH TOLL

A few thousand, primarily in North America and Europe. Industrial cleaners are not commonplace in Third World countries, limiting the potential toll. In addition, child fatalities by accidental poisoning have declined nearly sevenfold since the Poison Prevention Packaging Law was enacted in 1970.

NOTABLE VICTIM

In her book, *My Life Among the Serial Killers*, Dr. Helen Morrison describes a gruesome torturing of victim Larry Pearson by Robert Berdella. Berdella dabbed Pearson's eyes with Drano and injected his vocal chords with the drain cleaner to keep him from screaming.

HORROR FACTOR: 9 

The error of one's ways is apparent as soon as the lye hits tender flesh, but by then it's too late.

GRIM FACTS

- Mortuary scientists have shown that lye is an effective agent for dissolving bodies. Rather than being buried or cremated, cadavers can be reduced to a syrupy residue and flowed down the pipes.

- Though sodium hydroxide is not considered carcinogenic, cancer of the esophagus may develop years after esophageal tissue is damaged by exposure to the chemical.

- Statistics show that accidental pediatric poisonings tend to coincide with changes in a child's routines, such as during holidays, illnesses, moving, stressful times, and celebrations.

DROWNING AT SEA

Water, water everywhere, but not a drop to breathe.

HOW IT KILLS

You don't have to look far to find a body of water to drown in. People slip beneath the surface every day in pools, ponds, streams, lakes, and rivers. Even bathtubs and five-gallon buckets are potential killers, since any collection of water more than two inches deep is a drowning hazard when you're face-down in it. Every now and then, water even comes to get you in the form of a flood. But the open ocean is the big drink.

Fish absorb oxygen through their gills much like you absorb it through your lungs. Unlike your fishy friends, though, you can't absorb the oxygen (O_2) in water because it's bound to two hydrogen atoms (that's the H_2 in H_2O). Clean water can cure all kinds of ills when you drink it—but when you inhale it, you're headed for a watery end.

There are few more fun ways to spend a summer day than motorboating with friends on the open ocean. Miles from highways, and without the Coast Guard in sight, you knock back a few beers and let the 25-knot winds blow back your hair. But a deathly crack tells you the boat has hit something—hit it *hard*—and before you know it the craft is riding on edge, then capsizes. You're trapped underneath. It takes roughly three seconds to determine which is the way to the surface, yet it feels like you've spent half the air left in your lungs figuring that out. You try to make a fast break for the sunshine—but your ankle is entangled in a tow line and your swimming is jerked to a stop. If you'd taken a full gulp of air before going under you may have had a minute or more to survive. In this case, though, it's been less than thirty seconds and your lungs are burning to exhale the body's waste gas, carbon dioxide, and inhale oxygen-rich air.

The buildup of carbon dioxide is like poison in your bloodstream, and your brain sends out a red-alert message for the body to provide oxygen by breathing. You resist as long as you can, but finally the body's craving for air overpowers your keen desire not to inhale while underwater. Your lungs fill with water like a cheap cabin on the *Titanic*. Even if your lungs could absorb the oxygen in the water, the tiny sacs where oxygen is exchanged with blood have collapsed under the weight of water. Blackness begins creeping in at the corners of your vision. As your body yearns for O_2, your heart struggles to keep a steady beat, then goes into ventricular fibrillation and stops.

Your brain is still firing neurons but is helpless to inspire any action that could save your life. The darkness completely overcomes your vision now and, in the final scene, your day of boating fades to black.

KNOWN BY SCIENCE AS:

Fatal aquatic accident

MEDICAL CAUSE OF DEATH

Asphyxiation; ventricular fibrillation

TIME TO KILL

One to three minutes

HIGHEST RISK

Unattended children; drunks operating motorboats; fishermen

LETHALITY

Very high. Because the brain survives longer than the lungs or heart, a lucky few are saved via resuscitation and defibrillation.

KILLS PER ANNUM

In the United States, there are more than 3,500 unintentional drownings every year. Seven hundred more die of drowning or other causes in boating accidents.

HISTORIC DEATH TOLL

Millions if not billions worldwide. Among sailors alone, tens of thousands of boats have been lost at sea over the last millennium, taking their entire crews down with them.

NOTABLE VICTIM

On December 17, 1967, Australian prime minister Harold Holt went for a swim at a favorite beach. Holt was an avid water sportsman but his friends urged him not to enter waters where the rip tides were known to be dangerous. He swam out, disappeared from view, and never resurfaced. For years rumors circulated that Holt had faked his own death to be with a mistress, but in 2005—thirty-eight years after he disappeared—a state coroner ruled Holt's death an accidental drowning.

HORROR FACTOR: 8

One minute may seem like a mercifully short time for death to come, but it's a *long* minute.

GRIM FACTS

- Though most drowning victims die of the causes described above, about 10 percent have laryngospasms. Once water hits their larynx (voice box), their body's reflex is to squeeze shut the larynx to prevent water from entering the lungs. No water ever does—but neither does any air.

- Alcohol is a contributing factor in half of all drowning deaths among adolescent males. On the whole, males of all ages are four times more likely to drown than females.

- Experiments in "liquid breathing" suggest that humans could, theoretically, survive breathing an oxygen-rich liquid such as a perfluorocarbon. Don't go filling your pool with perfluorocarbons just yet; the body will still build up toxic levels of carbon dioxide.

DRUG OVERDOSE AT A ROCK SHOW

We're not trying to perpetuate any stereotypes here—with a little sniffing around, you can probably find as many substance abusers among a crowd of rock and rollers as you can at their mothers' tea parties. But for argument's sake, let's honor two-thirds of the "sex, drugs, and rock 'n roll" triumvirate. Here's a look at what happens when you get a VIP pass to life's after-show party.

HOW IT KILLS

There is a full ensemble of famous rockers playing that great gig in the sky, and many opportunities for you to join their band. You could go down in a private plane, get electrocuted on stage, drown in a swimming pool, or choke on some vomit. But if you're aiming to drop off the charts quickly and perhaps painlessly, nothing beats an old-fashioned rock-and-roll overdose.

Any number of individual drugs, and many fateful combinations of those drugs, have been used in the name of creativity and/or self-medication. You've got your amphetamines and cocaine to kick you up before the show; your Vicodin and Xanax to calm the nerves (thanks, Doctor Feelgood); and then your morphine, Percocet, or heroin for chilling out afterwards. If you are true rock royalty, you'll knock 'em all back with a swig of Jack Daniel's. Of course, drug tolerance plays

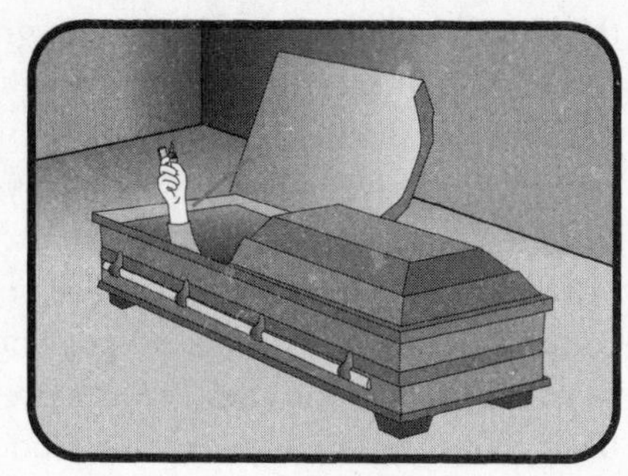

a roll in your ability to do the dope and live through it. Your tolerance can be increased by bumping up your drug intake, so you may want to blow an extra line here and there, or share a Pepsi with Keith Richards.

Stamping the word "overdose" on a death certificate tends to be an oversimplification because few drugs of abuse kill solely by means of their toxicity. Contrary to popular and past theories, even the purity of heroin is not believed to be a factor in most overdose fatalities; rather, drug impurities or the mixing of heroin with other drugs and alcohol is what gets you kicked by the horse.

If you're already feeling comfortably numb from a push of the needle, chances are good you will be in too much of a stupor to realize your breathing has become dangerously shallow and your heartbeat desperately slow. But to anyone watching, you're slipping into a coma. Your movement will be limited to an occasional muscle spasm, your breathing will be barely detectable, and your skin will have gone cold and clammy. Someone looking closely will note that your fingernails and lips are turning blue due to decreased blood circulation. Eventually, your breathing will slow to a stop. Respiratory failure will kill you on its own, though if that doesn't do it, your heart will fail from loss of blood pressure or your oxygen-starved brain will sing its swan song and die.

Since your heroin may have been sold to you under the street name "body bag" or "homicide," you really should've seen it coming. But maybe, instead, it's cocaine that makes you rock and roll all night. If heroin applies the brakes to your body's vital systems, coke stomps on the gas pedal.

Imagine the sun is coming up on your all-night binge. Having irritated and destroyed fragile tissue inside your nose from repeated intake of Peruvian marching powder, you've already been dabbing at the blood, and your jaw aches from grinding on your molars for the past eight hours. Now you snort up a sizeable snoutful. Just as you're about to tell another hilarious story to a room of admirers, you realize you don't feel so good. In fact, you suddenly feel awful. Your body temperature rises like you're in a sauna—or is it dropping? You can't tell; you're in a cold sweat. The powerful stimulant properties of cocaine are beginning to send your body into overdrive. With your heart racing in your chest, your extremities quiver and your entire body starts to convulse. As your blood pressure skyrockets, you are mere moments—perhaps split seconds—from a heart attack, stroke, or seizure.

And they say only the good die young.

KNOWN BY SCIENCE AS:
Substance overdose

MEDICAL CAUSE OF DEATH

- Respiratory failure; hypovolemic shock; brain death (heroin)
- Heart attack; stroke; respiratory failure (cocaine)

TIME TO KILL
Some overdoses are instantaneous while others take several hours. Research suggests that in more than half of all heroin fatalities, three hours or more pass between administration of the dose and time of death.

HIGHEST RISK
Frequent users partying alone. People who overdose are more likely to be saved if anyone nearby is sober enough to recognize they're in trouble and call for help.

LETHALITY

Medium, among regular users. Fatality rates are far higher among repeat users than first-timers.

KILLS PER ANNUM

About 17,000 people directly or indirectly from abusing illegal substances every year

HISTORIC DEATH TOLL

Undetermined, but includes dozens of notable musicians since the '60s and legions of their fans. The deaths of several high-profile artists in the early 1970s—Jimi Hendrix, Gram Parsons, Jim Morrison, Tommy Bolin—set a pace for the decades to follow.

NOTABLE VICTIM

The quintessential rock overdose goes to quintessential bad-ass Sid Vicious (né John Ritchie) of the Sex Pistols. In February 1979, the twenty-one-year-old Vicious scored some high-grade heroin the day after he was freed from prison, where he'd been drug-free for several weeks. He overdosed that night and woke up dead. For an anti-establishment icon, it was a pretty well-established way to go.

HORROR FACTOR: 3

Dying from an overdose has the built-in benefit of medication to ease the pain. Plus, history shows it can be a great career move.

GRIM FACTS

- Marijuana is as steady a rock-and-roll staple as "Brown Sugar," but there's never been a recorded case of anyone dying from a pot overdose.

- Keith Moon, the Who's drummer and resident maniac, OD'd on the pills he was taking to overcome alcoholism. He died in the same London flat where the body of "Mama" Cass Elliot of The Mamas and The Papas had been found four years earlier. (Contrary to pop mythology, the chubby Mama died of a heart attack in her sleep—not from choking on a ham sandwich.)

EBOLA VIRUS

The Ebola virus is Earth's very own Alien movie, except that the creature that eats its way through your internal organs is microscopic.

HOW IT KILLS

Medically speaking, the Ebola disease is known as Ebola hemorrhagic fever. Once contracted, symptoms include headaches, muscle and joint aches, abdominal pain, fever, fatigue, nausea, red eyes, and dizziness. Within a few days, new symptoms appear: bloody diarrhea, uncontrolled bloody vomiting, bleeding from the nose, bleeding from the mouth, and bleeding from the anus. All this is a result of the virus causing blood vessels to dissolve.

However, actual death from Ebola is usually due to organ failure or the lack of blood in the body. This latter condition is quaintly known as hypovolemic shock.

Because of how quickly it moves from host to host, no one is sure where the Ebola virus comes from, what causes outbreaks of the disease, or how it is initially transmitted to humans. Contact with fruit bats and infected primates are as good a guess as anyone has come up with.

Although no carrier source has been identified, transmission between humans occurs as a result of direct contact with the blood, secretions, organs, or other bodily fluids of infected persons. Once Ebola hits a village, it typically swarms through the inhabitants, infecting everyone it can until it runs out of hosts—and then disappears. Scientists have been unable to find where Ebola lives once an outbreak is over. In effect, it is the real-world slasher from bad teen-horror flicks, appearing out of nowhere to wreak carnage on unsuspecting citizens and then disappearing into the night once all the victims are piled up, leaving only blood in its wake.

And, yes, there is a lot of blood left in Ebola's wake.

KNOWN BY SCIENCE AS:

The Ebola virus, a member of the Filoviridae family of RNA viruses. First reported in 1976 in the Ebola River Valley of Zaire (now called the Democratic Republic of the Congo), the virus is believed to be zoonotic, meaning it is carried by animals. There are four types: Ebola-Zaire, Ebola-Sudan, Ebola-Ivory Coast, and Ebola-Reston.

MEDICAL CAUSE OF DEATH

Systemic multiorgan failure; hypovolemic shock

TIME TO KILL

Two to twenty-one days

HIGHEST RISK

Those most at risk of getting Ebola are hunters handling infected monkeys, medical personnel treating infected patients, and—how's this for irony?—people attending funerals of Ebola victims.

LETHALITY

Very high. There is no known treatment or cure for Ebola. Less than 20 percent of those who contract the disease survive, making it an almost certain ticket to the morgue. Since there is no cure, once you've got it, you have a better chance of getting elected president of the universe than surviving Ebola. Its lethality has led both the United States and the former Soviet Union to investigate Ebola's use as a biological weapon.

KILLS PER ANNUM

Forty to fifty worldwide

HISTORIC DEATH TOLL

More than 1,200 people have died from the various strains of Ebola. Deaths have been limited to the African nations of Democratic Republic of the Congo, Gabon, Sudan, the Ivory Coast, Uganda, and the Republic of the Congo.

NOTABLE VICTIMS

A forty-four-year-old Congolese teacher named Mabalo Lokela, who died on September 8, 1976, was the first human reported to contract the disease.

HORROR FACTOR: 10

At the top of the "horror meter" scale, compares to being burned alive or being eaten by a large predator

GRIM FACTS

- Contrary to popular myth, victims do not die from liquefaction of their internal organs.

- The Centers for Disease Control (CDC) reported that Aum Shinrikyo, the Japanese cult that used sarin gas to attack Tokyo's subway system, attempted to weaponize Ebola in 1992.

- Various primates have been known to contract the disease, notably Africa's lowland gorillas and chimpanzees, although antelopes and porcupines also get it.

Not-so-Hard to Swallow

Sherwood Anderson, noted writer and author of *Winesburg, Ohio,* swallowed a toothpick during a party, either from an hors d'oeuvre or a martini olive. Either way, he died months later on March 8, 1941, from the resultant peritonitis while onboard a ship sailing for South America.

EXECUTIONS

Yesterday and Today

Each nation has, throughout history, chosen a favorite way to dispense of dissidents, murderers, and traitors. From boiling and burning to garroting and gassing, capital punishment is the only way to get the government to stamp your death warrant "approved."

DRAWN AND QUARTERED

Long before the notion of cruel and unusual punishment gained any traction, the kings of England punished high treason by drawing and quartering. This method of execution was, and is, unparalleled in suffering and gore.

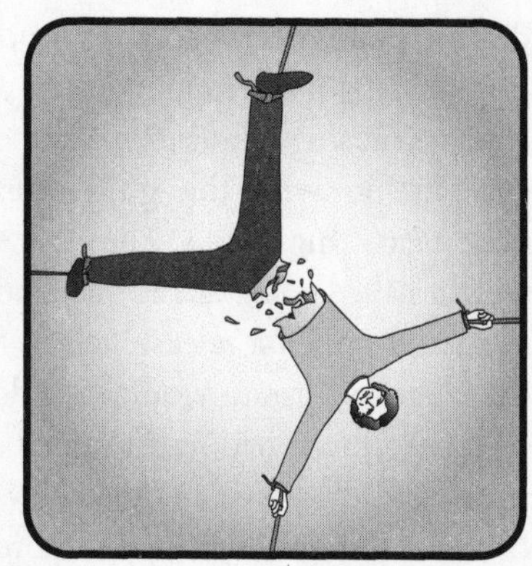

A condemned prisoner was first taken to the gallows and hanged, but cut down while still alive. If he lost consciousness, he was slapped and splashed with water so as not to miss the next stage, which began with being tied down to a block or "quartering table." There, his genitalia were removed before he was cut open and disemboweled. The punishers moved fast so that a prisoner would live to see his intestines and other organs spilled from his body—and then burned before his eyes. Once the body cavity was emptied, the prisoner was decapitated and his body cleaved into four pieces featuring one limb each.

Other than that, it wasn't too bad.

GRIM FACTS

- Historians debate whether the term "drawing" refers to the hanging stage, the strapping onto a quartering table, or the evisceration.

- In some instances, the method of quartering was to make partial cuts dividing the body, lash each limb to a horse, and spur the four horses.

- As a warning to others, the body parts and head were displayed throughout the country in separate locations of the king's choosing.

GUILLOTINE

Taken from the Latin *capitalis*, meaning "regarding the head," the term *capital punishment* has no more literal iteration than a beheading at the guillotine.

After the French adopted a law in 1791 that everyone condemned to death should die by decapitation, back-room architects of the French Revolution got to work on a beheading machine. Similar devices were already causing headaches around Europe, but the French perfected decapitation with the guillotine, whose gleaming feature was an oblique cutting blade. The 45-degree blade was bolted to a heavy weight called a *mouton* and suspended high on a crossbar. Once the executioner pulled a release (called a *déclic*) on the side of the guillotine, the blade and *mouton* would ride down rails and slice through the neck of a face-down prisoner trapped in a wooden collar. Skin, cartilage, spine, and the bones of the neck were no match for eighty-five pounds of *mouton* and blade dropping more than seven feet. With the neck severed by the angled blade in a fraction of a second, the prisoner's melon would thunk into a waiting bucket.

Can a decapitated head live long enough to detect a moment of pain, or catch a glimpse of the body from the basket? There are a number of claims to severed heads blinking, biting, moving their eyes, or winking at an executioner, but none is verifiable. It is true that the brain could briefly have some functioning after being separated from the rest of the nervous system, though it's expected the blow would cause immediate shock and unconsciousness. Without blood flowing nutrients and oxygen from body to head, brain death would be complete within fifteen to twenty seconds.

GRIM FACTS

- Dr. Joseph-Ignace Guillotin, a professor of anatomy, suggested the French develop a more humane method of execution during the Revolution, but was not involved in designing the device that bears his name.

- Beheading was the only method of execution in France for nearly 200 years. Guillotines stayed well greased until the last official execution in 1977.

LETHAL INJECTION AFTER LONG PRISON STAY

Among favored methods of execution, lethal injection is the category killer. It's used in 37 of the 38 states that have the death penalty. (Nebraska favors capital punishment but just can't pick a favorite.)

Redundancy is the name of the game in lethal injection, which might more accurately be called lethal injections. The procedure is intended to kill the prisoner three times over, with each deadly dose duplicated for backup.

The prisoner is strapped to a gurney and then connected to two intravenous lines, one in each arm. Then begins a series of intravenous drips. First is a saline solution, which in itself is harmless but which expedites a drug's rush to the bloodstream. Sodium thiopental is next down the line; a deliberate overdose of this anesthetic should render the prisoner unconscious and, it is presumed, into a state in which pain should not be detected. Saline flushes the line again, and then pancuronium bromide is injected to cause complete paralysis, which consequently causes respiratory arrest. But wait, there's more. The final solution, potassium chloride, is administered to stop the heart.

Death results from anesthetic overdose, respiratory failure, or cardiac arrest—take your pick. From first injection to death, the procedure usually takes just over eight minutes. When problems arise, though—thanks to collapsed veins, overly tightened straps, operator trouble—it can take more than twice as long. Following trouble with clogged tubes, it took the state of Illinois a full eighteen minutes to put down the Killer Clown, John Wayne Gacy.

GRIM FACTS

- In the United States, doctors do not participate in executions due to codes of medical ethics. Injections administered by poorly trained prison staff can leave undead prisoners groaning and thrashing on the table.

HANGING FROM THE GALLOWS

Sure, it seems archaic, but to this day death-row inmates in the United States still make their way to the gallows. The state of Washington, for example, favors lethal injection but will honor a request when a prisoner checks the box marked "hanging" instead.

The tried and true physics behind a hanging rely on a balance of

body weight and rope length. If the rope is too long for the prisoner's weight, his body gains too much force on the way down and he'll be decapitated; if it's too short, he'll dangle and strangle. It is an instantaneous breaking of the neck, rather, that makes for an ideal "long drop" hanging. The noose is placed around the neck of a bound prisoner with the rope's knot behind his left ear. When a trap door opens beneath his feet, the prisoner's falling weight creates a 1,000-pound force that snaps the spinal cord between the two topmost vertebrae, which attach the skull to the spine. Though the prisoner's head stays on (usually), he suffers the same effects as a decapitation.

Hanging has built-in fail-safes, too. If the neck isn't broken, the prisoner will still die of strangulation from a closed or shattered airway. Alternately, the rope's squeeze on the carotid arteries can result in lack of blood to the brain (cerebral ischemia).

GRIM FACTS

- One unsuccessful variation on hanging was the "upright jerk." Using a system of weights and pulleys, the neck of a victim was supposed to be broken as he or she was jerked into the air.

- Ropes used in hangings are boiled and stretched in advance to prevent a bungee-like spring effect.

THE GAS CHAMBER

If the afterlife has a waiting room, the gas chamber is it. And the wait is about ten minutes.

Gas chambers are akin to bathyspheres, complete with windows and an airtight door, though they're designed to keep deadly gases in rather than pressurized water out. A prisoner led into the chamber is seated and strapped into a chair before being sealed inside alone.

Directly under the chair is a bucket of sulfuric acid. On a signal from the prison warden, the executioner hits a switch, which drops crystals of sodium cyanide into the bucket. The reaction of the two chemicals causes deadly hydrogen cyanide gas to rise.

In the movies, spies taking cyanide pills die quick and painless deaths. The reality of a gas-chamber execution is not quite so tidy. Cyanide goes to work at the cellular level, preventing cells from getting the oxygen they need to keep the tissues of the body alive. A deep inhale of hydrogen cyanide may bring death on fast, but prisoners more commonly hold their breath or take shallow breaths to delay the inevitable. The struggling and anxiety-wracked prisoner experiences the sensation of a prolonged heart attack as the heart is deprived of oxygen. With millions of oxygen-starved cells dying, which kills ever more tissue, the pain increases and the prisoner begins to lose control of his faculties. The death of enough cells eventually causes the death of the system they support. Following seizures and anecdotal episodes of eye-rolling, drooling, defecating, and vomiting, the brain dies of hypoxia (lack of oxygen).

However, the heart still beats, and the prison's medical team may monitor a beating heart from outside the chamber for a good seven minutes before declaring the prisoner dead. The chamber air is then cleared, and orderlies recover the body about half an hour later. A final action is to ruffle the dead prisoner's hair—a macabre "atta boy"—to release any vestigial traces of the gas.

GRIM FACTS

- The gas chamber was developed after prison authorities tried to off Nevada inmate Gee John by pumping cyanide gas into his cell while he slept. The gas dissipated and John survived. John later became the first person to die in a gas chamber.

ELECTRIC CHAIR

A prison is a miserable place to be, and the electric chair is easily the worst seat in the house. Prison staff routinely undergo psychological evaluations before and after executions, and witnesses are known to faint, vomit, and suffer nightmares for decades. But they get to walk away.

In preparation for execution, a prisoner is usually shaved to reduce electric resistance before being belted to the chair. A sponge dampened with saline solution is placed on the head under a metal electrode that caps the prisoner's scalp and forehead. Electro-Creme, a conductive jelly, is also spread over the scalp and over another electrode that attaches to the leg. The throw of a switch then subjects the blindfolded prisoner to as much as 2,350 volts at over 7.5 amperes of alternating current—more than one hundred times the power needed to run a microwave oven. After a jolt, which can last anywhere from five to thirty seconds, the execution team has to wait for the prisoner's body to cool down before checking for a pulse. If the heart is still beating, the prisoner is hit with another surge. He may die in the first few moments, but if not, the process continues—jolt and rest, jolt and rest—until he's dead.

The exact cause of death from a sit-down with Ol' Sparky has been the subject of debate. Some pathologists have cited the "massive depolarization" of the brain from the surge of electricity, which would cause instantaneous brain death. Even if the brain survived the shock, they say, a rise in brain temperature upwards of 145 degrees Fahrenheit would be enough to kill the organ. Others says heart failure kills before brain death; the electric shocks superfibrillate the heart, causing it to race far beyond its capacity or cease beating after myocardial infarction. Still others describe that the shock and heat coursing through a body causes abnormal contraction of muscles and internal organs, resulting in asphyxia by contracting the larynx or the lungs.

The electric chair was first built for the state of New York, which sought a humane method to replace hanging. Nice thought, but the stories of botched electric-chair executions are as grisly as the crimes that put prisoners on the hot seat in the first place. Witnesses have cited eyeballs popping out of the head, foaming mouths, flesh turning red and bursting, heads on fire (though it's sometimes the sponge that first catches fire), and the "sickly sweet smell of burning flesh." Sounds from the execution room have been likened to bacon frying.

GRIM FACTS

- Electrocution is no longer the preferred method of execution anywhere in the United States, but nine states allow it as an alternative to lethal injection.

FALLING FROM A MOUNTAINTOP

The thrill of a technical climb is in cheating gravity and fate. Here's what happens when the debt is collected.

HOW IT KILLS

There are several ways to meet disaster on a mountaineering expedition or even on a simple day climb. Among the many risks are avalanche, exposure, cerebral edema, pulmonary edema, lightning, and "trundling," where an unsuspecting climber is clobbered by a stone thrown from an equally unsuspecting tosser at the top.

Falling, though, is the fear humans are naturally prewired for, and justifiably so. A tumble down twenty-five feet of talus can easily be as deadly as losing one's footing on a mile-high ridge. The fall needn't be especially steep to kill you, as climbers on freehand scrambles have been known to slide and cartwheel uncontrollably before laying still and broken on the rocks below. You may even die well before your body comes to a stop.

Attempting a technical climb unroped (freestyle) is to some the mark of expertise, and to others the mark of stupidity. Even with a rope and years of experience, climbers meet their end every year due to handholds breaking off, chocks or ice screws pulling out, inadequate belaying, poor anchoring technique, being shucked off a rope by another falling climber, and equipment failure. Those not killed immediately by a fall's impact may perish after laying injured in a crevasse, on a remote ledge, or in any area where access to proper care involves a time-consuming carryout. Periodically, on high-altitude expeditions, a climber lands mere feet from a route but can't be helped down the mountain since assisting the incapacitated climber would endanger the lives of rescuers. On some of the world's highest peaks, the irretrievable bodies of long-ago fallen climbers are part of the scenery.

KNOWN BY SCIENCE AS:

Inability to self-arrest (meaning "to stop one's self")

MEDICAL CAUSE OF DEATH

Blunt force trauma to head; spinal injury leading to asphyxiation. Secondary causes include complications from untreated injuries and, in some climates, exposure when a fallen climber cannot be evacuated.

TIME TO KILL

Instantaneous upon landing; one to two days by secondary causes

HIGHEST RISK

Experienced mountaineers climbing unroped on rock

LETHALITY

High. At any height, a fall onto rock can result in a fatal head injury.

KILLS PER ANNUM

On North American climbs alone, an average of thirty climbers die every year.

HISTORIC DEATH TOLL

In the 6,085 mountaineering accidents reported in the United States over the past fifty years, 1,352 people have died. That is, more than one in six U.S. accidents has involved a fatality.

NOTABLE VICTIM

Alison Hargreaves, the first woman to summit K2, was blown off the summit ridge during her descent. Her body was never recovered. Some believe the 28,251-foot mountain is cursed for female climbers: Between 1986 and 2004, the only five women to summit K2 either died on the way down or perished on other peaks soon afterward.

HORROR FACTOR: 8

Falling to one's death is the realization of a universal fear: this time, you're not going to wake up before you hit. In a free fall there is no pain to distract one from imagining the inevitable, body-breaking end.

GRIM FACTS

- Annapurna is considered the world's deadliest mountain, though this Himalayan peak is only the world's 10th highest.

- Staying tethered to a rope is no guarantee of surviving a fall. If a climbing rope is passed under a leg, the climber may be jerked upside down at the end of the slack length, increasing the risk of being slammed against a stone face while dangling from the line.

- You needn't travel to the Himalayas to die falling off a mountain. Alaska's Mount McKinley, aka Denali, claims the life of one in every 100 climbers.

FATAL INSOMNIA

The idea that you can be "dead tired" is familiar to all of us. But fatal insomnia is a disease you will lose sleep over, because you'll die before you ever get another good night's sleep.

HOW IT KILLS

Fatal insomnia is exactly what it sounds like, a disease that kills you by preventing you from sleeping. The gene that causes fatal insomnia is believed to trigger bizarre mutations in brain proteins called prions. These prions are thought to become infectious and start devouring the thalamus, the portion of the brain that controls sleep. They eat entire sections of nerves, leaving holes like those found in sponges.

The disease, which has been found in about three dozen families worldwide, strikes in middle age, and begins with signs of mental breakdown. You will have panic attacks related to nothing at all, along with chronic insomnia. You may not be able to sleep for days at a time. This is the signal that you're in for a year and a half of sleep-deprived hell.

After about four months of feeling miserable from lack of sleep, you'll start sweating heavily all day long and bumping into things as you stumble about. Panic attacks will get worse, your heart will race, you'll be constantly irritated by everything around you, and you'll start to hallucinate while you're awake. This will last for about six months.

Eventually, everything devolves into pure insomnia. You simply will not be able to sleep regardless of your exhaustion level. You'll also have lost significant weight by this time, and you may not be able to control your bladder. Expect this to go on for roughly three months.

You have half a year to live, and it's going to be the worst six months of your life. You will have a complete mental breakdown as your life is lived in an ongoing hallucination. Your body will twitch uncontrollably, and you will howl in pain as your body tries to find some relief from its inability to sleep. Eventually, you will become unable to speak, unable to walk, and will fall into a coma. Death will happen very suddenly, but not suddenly enough.

KNOWN BY SCIENCE AS:

Fatal familial insomnia. Specifically, FFI is a transmissible spongiform encephalopathy (TSE), and is a hereditary prion disorder resulting from an autosomal dominant mutation.

MEDICAL CAUSE OF DEATH

Heart failure; respiratory failure; brain damage

TIME TO KILL

Approximately eighteen months after first symptoms appear, and the disease usually strikes around age fifty.

HIGHEST RISK

Those whose families carry the gene that predisposes them to dying from FFI

LETHALITY

High. If you have the gene for FFI, you are going to die from it.

KILLS PER ANNUM 

One or two

HISTORIC DEATH TOLL

Fewer than one hundred

NOTABLE VICTIM

In 1984, an Italian man known as "Uncle Silvano" allowed himself to be filmed in a lab during the final phases of his battle with fatal insomnia. After dying a horrific death, his brain was removed and provided the first insights into what causes FFI.

HORROR FACTOR: 2

You'll know that it's a possibility once you realize you've been born into the wrong family, but you won't know that you've got it until you start losing sleep over it.

GRIM FACTS

- The deviant prions that cause fatal insomnia are also thought to be at the root of "mad cow" disease and kuru, a disease found among cannibals in New Guinea.

- Odds for the average person getting fatal insomnia are one in 33 million. If it's part of your family DNA, however, the odds are closer to fifty-fifty.

- Stanley Prusiner used brain material from FFI victims as part of the research that won him a Nobel Prize in 1997.

The Pen Is Not Mightier than the Sword

On July 7, 1982, Olympic gold medalist Vladimir Smirnov was competing in a fencing tournament against Matthias Behr of West Germany. During a particularly forceful lunge, Behr's blade snapped and pierced Smirnov's face mask. The sheared-off blade drove through Smirnov's eye and into his brain. Though he was "technically" killed almost instantly, Smirnov was kept on life support for nine days.

FREEZING

Bears have fur, whales have blubber, and people have Gore-Tex. Without protection from the cold, hypothermia kills. Like a chicken breast tossed in an ice box, we freeze gradually from the outside in.

HOW IT KILLS

Stick a thermometer in a healthy person's mouth and the mercury should read 98.6 degrees Fahrenheit. Everyone's familiar with body temperature rising to fight infection, but it's when your core temperature starts dropping that you really need to worry.

When the body can't generate as much warmth as it loses, hypothermia gradually starts to settle in. The most common causes of hypothermia are prolonged exposure to cold air; exposure to rain without the capability to dry off; and being dropped into water. Water draws heat out of the body faster than air does, and you don't need to fall into freezing waters to get hypothermia and die—given enough time in the drink, any cool dunk will do.

As your body temperature drops a mere two degrees below normal, the body will undertake several autonomic responses to warm itself: You'll start shivering in an attempt to increase movement and blood flow, and your extremities will become increasingly numb as the blood vessels constrict to reduce heat loss. Next, your breathing will become shallow. At a core body temperature of 95 degrees Fahrenheit, you'll officially have hypothermia, and your body will begin sacrificing heat needed at the extremities in an attempt to keep vital organs warm. Your skin will become cold and pale, your fingertips, toes, lips, and nose may turn blue, and shivering will become more severe. You'll start to lose muscle coordination, which will be marked initially by the loss of small-motor skills and then by out-and-out stumbling. Confusion and disorientation will set in and you'll slur and mumble incoherently. All of your body movements will become noticeably slower. Finally, unable to cope with the heat deficit, your heart or respiratory system will seize entirely.

An oft-cited phenomenon is that hypothermia victims will remove their clothing at a time when body coverage is fundamental to survival. Between 20 and 50 percent of hypothermia victims are found partially or completely undressed. "Paradoxical undressing" is the result of peripheral vasodilation, which is the dilation of blood vessels resulting from the body's attempt to redistribute its remaining heat. Vasodilation affects the sensation of warmth, just like when your face goes red from embarrassment. Though body temperature is plummeting perilously, the victim mistakes the skin sensation for body temperature, and undresses to cool down.

In cases where the lowering of body temperature is gradual, another curious behavior is that people will seek or create small spaces in which to curl up. Such victims are found in cupboards, under beds, or in a tiny snow cave. This phenomenon is known as terminal burrowing, indicating it's the last hole you'll ever dig. One more, however, will be dug for you.

KNOWN BY SCIENCE AS:
Lethal hypothermia

MEDICAL CAUSE OF DEATH

Cardiac or respiratory arrest

TIME TO KILL

In cold waters (say, 45 degrees Fahrenheit), one to three hours

HIGHEST RISK

Homeless alcoholics; people exposed to elements following a disaster; lost winter hikers

LETHALITY

Medium. The body can withstand several stages of hypothermia before the condition kills.

KILLS PER ANNUM

More than 700 deaths in the United States

HISTORIC DEATH TOLL

More than 75,000 in the twentieth century, though hypothermia is historically underreported in water fatalities. It's difficult to swim with no muscle coordination, but hypothermia only gets credited with an assist when drowning is found to be the cause of death.

NOTABLE VICTIMS

Things went from bad to worse for the infamous Donner Party back in the winter of 1846–47. When the California-bound settlers were caught in a blizzard, seven died of exposure and were cannibalized by other members of the group. Some party.

HORROR FACTOR: 7

Hypothermia can take hours to kill, leaving the victim plenty of time to dream of sandy beaches and think of all those times an air conditioner seemed like a great idea.

GRIM FACTS

- Hypothermia is the Number 1 killer of outdoor recreationists.

- According to physicians at Mayo Clinic, early stages of hypothermia are marked by the "umbles": stumbles, mumbles, fumbles, and grumbles.

- Brain death by lethal hypothermia is rare. The brain may continue to function more than an hour after a victim loses consciousness.

GANGRENE

It sounds bad, and it smells even worse.

HOW IT KILLS

There are two types of gangrene, both of which are associated with dead or dying body parts. Dry gangrene is caused when blood flow to tissue is cut off, often because of blocked arteries. Wet gangrene is caused when an infected wound cuts off blood supply, and this is the kind of gangrene you really don't want to get. One form of wet gangrene is called gas gangrene, which occurs when *Clostridium perfringens* bacteria invades a wound and begins producing toxic gas.

Let's imagine that you've received a nasty gash on your leg while digging in your garden and decide it's not worth cleaning or treating. Anaerobic bacteria which has been laying dormant in the soil finds a nice moist breeding ground inside your flesh, where it thrives in a near-airless environment. The bacteria infects the area around the gash, from skin to nearby tissue, muscle, and organs. These all swell from the infection, and start to cut off blood flow. Trapped and stagnant blood provides even more fuel for bacterial growth. And

a colorful side effect emerges: your flesh starts to turn black and even blue as it starts rotting.

As the bacteria reproduces, it creates a gas (made up mostly of nitrogen) with a distinctly sickly sweet odor, which makes the whole process of having gangrene that much more grim. This gas, seething within your body—you can feel the gas bubbles under your flesh—quickly spreads to other organs and tissue. As they swell and are cut off from blood, they begin to die, or "necrotize" as the emergency room doctors like to say. The bacterial toxin from these areas is eventually absorbed into other parts of the body, causing septicemia.

Vomiting then commences, followed by increased heart rate and breathing. Low blood pressure and coma are the last stages of the infection, which leads to death.

KNOWN BY SCIENCE AS:

Gangrene is a general medical term; gas gangrene is known as clostridial myonecrosis. The condition is an infection by *Clostridium perfringens* bacteria.

MEDICAL CAUSE OF DEATH

Septicemia; restricted blood flow; shock

TIME TO KILL

Depending on the severity of the wound and the resultant bacterial infection, death can occur within forty-eight hours or drag on for several weeks.

HIGHEST RISK

Diabetics; people with leukemia and colon cancer; people with blocked blood vessels; battlefield soldiers; patients who have undergone surgery in less-than-sanitary conditions

LETHALITY

High. Undetected, gangrene has close to a 100 percent kill rate. If detected early enough, however, gangrene can be treated, although amputation of the infected area is one of the common cures.

KILLS PER ANNUM

Dozens worldwide, mostly victims of Third World military conflicts

HISTORIC DEATH TOLL

Tens of thousands. Significant numbers of the battle-related deaths during the American Civil War and World War I were a direct result of gangrene from poorly treated wounds, and this condition persists in Third World wars.

NOTABLE VICTIMS

- Tutankhamun, the famed boy king of Egypt who died at age nineteen around 1323 B.C., was thought to have died from gangrene after breaking his left leg. This theory is supported by results of a CAT scan conducted on King Tut's corpse in 2005.

- Allan Pinkerton, founder of the Pinkerton Detective Agency and head of the organization that became the Secret Service, slipped while walking on a Chicago street and bit his tongue. The wound developed gangrene, and Pinkerton died three weeks later on July 1, 1884.

- Herod the Great, scourge of the ancient Jews, may have died from a rare form of gangrene in his genitals, according to research presented at the 2002 Clinical Pathologic Conference.

HORROR FACTOR: 10 

Watching your skin turn black and emit odors like an open sewer is no way to spend your last hours.

GRIM FACTS

- The anaerobic bacteria that causes gangrene is related to the bacteria that causes botulism, tetanus, and toxic shock.
- The word "gangrene" comes from the Greek word *gangraina*, which describes a festering sore.

GARROTE

One of the quietest methods of killing involves pulling a wire across your throat. No yelling, no noisy guns. Most of all, no mess.

HOW IT KILLS

Garroting is the process of strangulation using a thin line, such as a wire, to choke the life out of a victim. It is, perhaps, the ultimate stealth weapon.

Once upon a time, garroting was a means of execution, but that doesn't happen much anymore (although Spain practiced it up until the 1970s). Beyond that, it has always been a favored method of dispatching humans quickly by those who need to keep their killing nice and quiet. This includes assassins, spies, serial killers, and perverted stalkers.

The whole process of garroting is designed to make it easy on the assassin and quick on the victim. Which, in the latter case, is you. Your assailant has a line a foot or so long—it can be a guitar string, nylon fishing line, picture wire, or some other generally unbreakable line—and has attached each end to a handle, such as a wooden dowel or a stick. The wire is pulled around your neck from behind, so that the assailant can yank backward on it. This compresses your trachea and pinches it shut. You can't get air past the point of closure, so you can't breathe.

At the same time, your larynx—also known as your voice box—is squeezed shut, rendering you unable to scream out for help. The assailant turns the handles of the garrote to twist it tighter against your throat (all the while, the handles are protecting his own hands from having the wire cut into them). You will quickly pass out from lack of oxygen while trying to find a way to loosen the garrote from your throat. You will not be able to do this because of the nature of the garrote, which is essentially a slipknot around your neck that allows for no slack. Meanwhile, the wire may actually cut into your flesh, causing you to bleed and rupturing your trachea.

As the wire continues to block any air from entering your body, you will die from asphyxiation—and it will happen without you uttering a word.

KNOWN BY SCIENCE AS:

Strangulation

MEDICAL CAUSE OF DEATH

Asphyxiation

TIME TO KILL

This is a function of how long you can hold your breath in a desperate situation. Three minutes, max.

HIGHEST RISK

Secret agents; Spanish criminals

LETHALITY

High. Once it starts, the person garrotting you is not likely to finish until you are.

KILLS PER ANNUM

Best estimates are several dozen

HISTORIC DEATH TOLL

Thousands, as it has been employed by the Romans, Spanish, and Portuguese as a method of capital punishment for nearly 2,000 years

NOTABLE VICTIM

JonBenet Ramsey was killed by a makeshift garrote in her home on December 25, 1996. The six-year-old beauty pageant queen was found dead with a garrote made from nylon rope and a paintbrush handle around her neck.

HORROR FACTOR: 10

As soon as you know you're being strangled by somebody who intends to kill you, the horror factor jumps all the way to the top.

GRIM FACTS

- The Spanish created special garroting chairs for executions. Some versions used screws to tighten the rope around the condemned's neck, while other chairs substituted an iron collar that could be tightened for the rope or wire.

- The word "garrote" comes from the Spanish word for cudgel or club, and actually refers to the wooden handles used as part of the strangulation device.

- The infamous Indian murderers of the seventeenth century, known as "Thuggees," killed unwitting travelers with garrotes made of yellow cloth.

- Canadian Major John Richardson wrote in his *Journal of the Movements of the British Legion* (second edition, 1837) that "I have no hesitation in pronouncing death by the garrote, at once the most manly, and the least offensive to the eye."

The Bigger They Are

Professional wrestling champ Owen Hart was entering the ring for the Intercontinental Championship on May 23, 1999...from the ceiling of the Kemper Arena in Kansas City. While being lowered on a cable, the release on his harness malfunctioned and Hart dropped seventy-eight feet to the floor, dying of trauma in front of thousands of wrestling fans. The announcer had to repeatedly tell the crowd and home viewers that the fatal fall was not part of the act.

GOING OVER NIAGARA FALLS

What would compel someone to plunge down Niagara Falls in a barrel, a ball, or a boat? There's a perfectly nice view from shore.

HOW IT KILLS

Ever since 1901, when schoolmarm Annie Edson Taylor survived the 170-foot plunge in a 54-inch oak barrel, daredevils have periodically cast themselves over the brink of Niagara Falls. Taylor, who was age sixty-three when she went over, later said she'd sooner "walk up the mouth of a cannon, knowing it was going to blow me to pieces" than take the ride again. Despite her admonition, not to mention the physics at play, another sixteen ultimate thrill-seekers have gone over Niagara in the past century. Some, including an unhelmeted clown in a kayak and an ill-prepared jet skier (see below), have met their end in the churning waters.

Surviving the free fall is no small feat, but the drop itself represents only a portion of the risk. Water cascades over the cataract at a force of

600,000 gallons per second, battering down on a vessel after it lands and tossing around its occupant like a rag doll. The potential for severe injury is high as a barrel is pounded by water or smashed against rocks after an otherwise safe landing. Barrels may also get stuck behind the thundering curtain of water, and without being spit out into open water where one can be freed from a barrel, riders suffocate inside.

Living through a ride over the Falls appears to be a matter of planning and pure luck. One theory is that cones of highly pressured water sometimes cushion the fall. Padding the inside of a vessel with inflated pillows or mattresses similarly slows the acceleration of a body as it slams into the container's wall. Another theory is that survivors don't actually plunge but instead ride down the falls like a body surfer, encountering a steep slope rather than a flat landing. That may explain how Kirk Jones survived his completely *al fresco* drop in 2003, with no barrel, no life jacket, and no protection whatsoever.

KNOWN BY SCIENCE AS:

Stunting without a license

MEDICAL CAUSE OF DEATH

Suffocation; blunt force trauma

TIME TO KILL

May be instantaneous with a bad landing on water or rock; several hours in cases of suffocation

HIGHEST RISK

Equipment failure, poor or heavy barrel design. Instances of suffocation are lower with supplemental oxygen onboard.

LETHALITY

Low, considering the activity. Eleven of the seventeen people to go over (which includes two men who repeated the ride and two dual-passenger descents) have survived.

KILLS PER ANNUM

Barely 1. Attempts over the past century average only about one every six years.

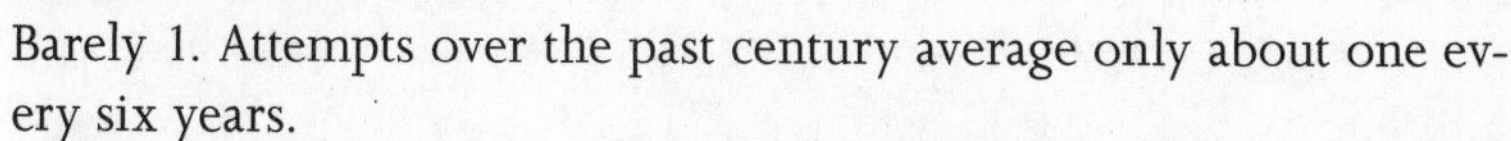

HISTORIC DEATH TOLL

Five humans and several animals. Barrel designs have improved, increasing the likelihood of survival, since Annie Taylor went over the brink in her wooden cask. However, the desire among daredevils to achieve new "firsts"—first without a helmet, first on a jet ski—is swaying the death toll back in nature's favor.

NOTABLE VICTIM

In 1995, thirty-nine-year-old Robert Overacker rode a jet ski over the Canadian Horseshoe Falls, reportedly to promote the plight of the homeless. His plan was to deploy a rocket-propelled parachute strapped to his back, let the jet ski crash below, and float gently into a pool at the base of the falls. According to the Niagara Falls Daredevil Museum, "his parachute did not open and Robert ended up promoting better parachutes."

HORROR FACTOR: 2

We can't ascribe a significant level of horror to a final fall over Niagara when the victims essentially sent fate an engraved invitation. You asked for it, you got it.

GRIM FACTS

- Stunters have only gone over the Canadian "Horseshoe" Falls, not the American Falls, which are lined with rocks at the bottom.

- The very first "stunt" at Niagara was a publicity stunt. In 1827, three local hotel owners sent a condemned schooner loaded with animals over the edge. Two bears jumped ship before the drop. A buffalo, two raccoons, and a dog all perished, though the goose lived.

- Niagara's animal history became a legacy, of sorts. Annie Taylor reportedly sent a cat down to test the waters; the cat died, though Taylor still made her famous ride. In 1930, George Stathakis brought along his 100-year-old pet turtle, Sonny Boy, as a good luck charm. Their 2,000-pound barrel was trapped behind the curtain for eighteen hours. Sonny Boy was indeed lucky, but Stathakis suffocated.

- Circus stuntman Bobby Leach went over in 1911. He suffered a broken jaw and two busted kneecaps, but survived. Fifteen years later he died of complications from a fractured leg after slipping on an orange peel.

GREAT WHITE SHARK

The great white shark has been assigned the role of most evil killing machine on the planet. It is arguably the most feared animal the world over, simply because the only way to be completely safe from it is to never go in the ocean—ever again.

HOW IT KILLS

The great white can grow up to just over twenty feet and approximately three tons. It is a steel gray color, except for its bright white underbelly. Found primarily in the Atlantic and Pacific oceans, the territory of an individual great white can be thousands of miles in any direction.

Most of the great white's life is spent looking for prey, which consists primarily of seals, large fish, dolphins, and the occasional surfer. It is attracted to vibrations in the water, as well as blood, and finds food using the oddly named ampullae of Lorenzini, a sensory organ that detects electromagnetic fields in water from miles away.

The great white is a fast and strong swimmer, estimated to reach speeds of 35 miles per hour or more. The fastest human swimmer (and you are not that human) has reached speeds of roughly 5 mph. Suffice it to say that the great white swims roughly twenty times faster than you do. The great white moves so quickly that it requires little stealth to attack its prey. Typically rushing up from below, the great white hits you with the force of an out-of-control pickup truck hitting a pedestrian.

The shark's jaws open while its eyes roll back into its head (presumably to keep them from being damaged during the attack). The top jaw juts forward out of the head and scissors down over the lower jaw. Several hundred teeth, each of which is serrated and up to three inches in length, clamp down on you with an estimated force of three tons per square centimeter. The shark then shakes its head like a pitbull tearing at a chew toy, and rips off whatever is in its mouth. This is typically something large, like an arm, leg, or entire torso. The shark devours this chunk without chewing, then returns to your body, weakened by blood loss, to rip another morsel. By the time the shark is done, there is little left to sustain life, including tissue, vital organs, flesh, bone, and blood.

KNOWN BY SCIENCE AS:

Carcharodon carcharias, from the Greek words for "sharp tooth." A member of the lamnidae family, it is the largest carnivorous shark.

MEDICAL CAUSE OF DEATH

Blood loss; organ failure; trauma

TIME TO KILL

Less than three minutes

HIGHEST RISK

Surfers in Australia, South Africa, and the California coast; swimmers with blood-flow problems; water-bound hemophiliacs

---

LETHALITY

Surprisingly low. Only about an eighth of all attacks on humans end in death. This is due to the shark ceasing its attack after removing a limb (a common occurrence) or realizing that humans are not as appetizing or as passive as seals. Several survivors have punched the nose of the sharks to stop attacks.

KILLS PER ANNUM 

One

HISTORIC DEATH TOLL

Records indicate that sixty-three people have been killed by great whites, while several hundred have been maimed and mutilated.

NOTABLE VICTIM

On December 19, 1981, Lewis Boren went surfing. His surfboard washed ashore the next day with a huge great white shark bite out of it. When Boren's body was found a week later, the chunk missing from his body exactly matched that of the surfboard piece.

HORROR FACTOR: 12 

To be eaten by a creature that is bigger than a minivan, and to have it done in stages, piece by piece, fully knowing that you are ending your life merely as food for a marine animal has to be so horrific that it defies any combination of words you can put together in the English language.

GRIM FACTS

- The contents of great white shark stomachs have been known to include boat motors, sheets of metal, other sharks, and entire unchewed seals.

- Scientific evidence suggests that a great white shark can be repelled by applying pressure to its snout, where much of its sensory system is believed to be concentrated in an area known as the ampullae of Lorenzini.

- The great white shark remains the only apex predator never captured and kept alive for an extended period by humans.

- Of nearly 400 species of sharks, only a handful are classified as maneaters, including the great white shark, the tiger shark, the bull shark, the mako shark, and the hammerhead shark.

GUINEA WORM

There is not another organism in this book that kills with such pure science fiction–style horror as the guinea worm. For something that is all but invisible when it gets inside you, it certainly leaves a big impression when it comes out.

HOW IT KILLS

The guinea worm lays its eggs in ponds and water supplies throughout central Africa. Once deposited in the water, the eggs are eaten by tiny fleas, which are so small they get swallowed with drinking water. Once inside your stomach, your digestive juices dissolve the fleas, but not the worm eggs inside them. The now-exposed eggs then nestle inside your intestines. Living comfortably, they remain in place until they grow to about three feet long. This takes about a year. At maturity, female worms mate with males and create new eggs of their own. However, they need to lay these eggs in water, and start the slow process of burrowing through your intestines and organs in order to get to your skin.

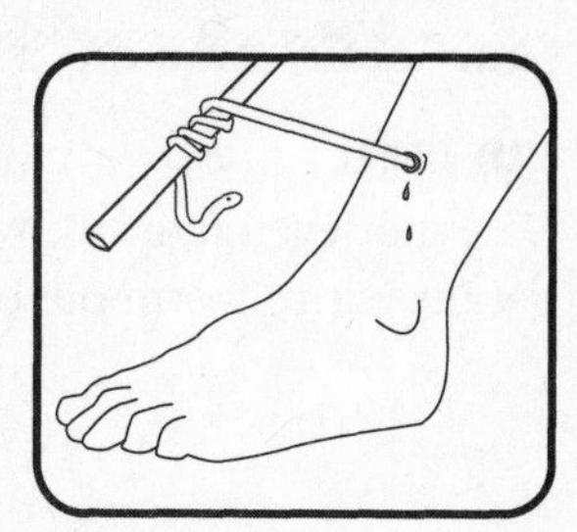

Once the worms reach your skin, or an eyeball—but before they break through to the outside—they release an acid that causes your skin to blister and bubble in order to create an exit route out of your body. These blisters are extremely painful, and as you try to ease the burn by washing them with water, the guinea worm bursts out of your body in a single, spaghetti-like strand.

This won't kill you, yet. As soon as it pops out of your body, you have to grab it and hold onto it. Then you begin wrapping it around a stick or something else that you can carry around with you. Over the course of the next several weeks, you have to gradually pull the entire worm out of your flesh, inch by excruciating inch, all the while making sure it doesn't break. If it does, what remains in your body can calcify and create hard bony spurs under your flesh.

As if all this wan't bad enough, and we think it is, what ultimately kills you is not the worm itself, but sepsis or tetanus caused by secondary infections of the worm's exit wounds. Yes, *wounds*, plural. Typical sufferers have an average of two worms coming out at the same time.

KNOWN BY SCIENCE AS:

The nematode (worm) *Dranculus medinensis* causes the infection known as Dracunculiasis. The term *Dranculus* comes from the Latin for "little dragon."

MEDICAL CAUSE OF DEATH

Sepsis; tetanus

TIME TO KILL

Weeks or even months, which is the waiting time to make sure you get the entire worm out of your system

HIGHEST RISK

Residents who use open water systems in the following countries: Benin, Burkina Faso, Ivory Coast, Ethiopia, Ghana, Nigeria, Mali, Mauritania, Niger, Sudan, Togo, Uganda

LETHALITY

Minimal. Mortality associated with the worm is low due to long traditions of handling it once it emerges from the skin. In addition, eradication efforts have dropped the number of cases from several million a year in the 1980s to around 10,000 per year today.

KILLS PER ANNUM

No one knows. Because of the remote location of most cases, there is little statistical information on mortality rates. Our estimate is less than a dozen.

HISTORIC DEATH TOLL

Again, no one knows, but is likely to be in the hundreds if not thousands.

NOTABLE VICTIMS

In 2002, two volunteers of the Nigerian eradication effort had their guinea worms removed using a traditional folk method of burning the worm. They died when they contracted tetanus after the "operation."

HORROR FACTOR: 9

Hoping that you can survive pulling the worm out over several months, and then not get the wound infected with a life-threatening disease, has got to make you worry whether things could get any worse.

GRIM FACTS

- If a guinea worm successfully erupts in open water, it lays millions of eggs, which starts the entire cycle over again.

- Many African villagers will not allow their ponds to be treated with insecticides that will kill the guinea worm because the ponds are sacred.

- Former U.S. president Jimmy Carter led a drive to eradicate Dracunculiasis from the planet.

HAIR DRYER IN THE BATHTUB

Hair dryers are a long-running favorite of suicides who aim to end their lives among the soap and bubbles.

HOW IT KILLS

The human body is an unfortunately effective conductor of electricity. While some tissues like bone and fat have higher resistance, nerves and blood vessels are like the body's very own copper wiring. Dry skin can actually protect the body from conduction, but when wet, skin can pass more current through to vital organs.

Water itself is not an especially good conductor, but impurities in the water such as naturally occurring salts and minerals are just great at helping to complete a circuit. When a plugged-in hair dryer is dropped into a bathtub, the alternating current (which, in most households, flows in the ballpark of 200 volts) instantaneously hits the skin everywhere it meets the water's surface and travels through the body. The nature of AC is to flow in cycles, so rather than delivering a single zap it causes continuous electrocution until a circuit is tripped or blows.

On contact, the electrical energy converts to thermal energy, leaving a burn pattern on your body that matches where waves of water touched your skin. The cycling AC flow paralyzes you, spelling big trouble for your respiratory and cardiovascular systems. The paralysis of respiratory muscles causes apnea, a suspension of breathing. In the heart, electrical shock disrupts or entirely seizes the natural electrical activity that keeps the heart beating steadily. Even if this disruption—known as ventricular fibrillation—doesn't stop your ticker from ticking, tissues in your heart can still be fatally damaged, or your arteries may spasm and fail.

Electrical shock to the central nervous system, which governs all your body's systems, can simultaneously shut down both respiration and heart function in one fell swoop of cardiorespiratory arrest.

KNOWN BY SCIENCE AS:
Electrocution by household appliance

MEDICAL CAUSE OF DEATH
Cardiac arrest

TIME TO KILL
About three seconds

HIGHEST RISK
Suicides on a budget

LETHALITY
Very high

KILLS PER ANNUM
Five

HISTORIC DEATH TOLL

In the 1980s, before the Consumer Product Safety Commission recommended new standards for hair dryers, there were about eighteen electrocutions by hair dryer every year.

NOTABLE *ALMOST* VICTIM

In 2004, thirty-four-year-old William Wolfe of Texas drew a bath for his wife and surrounded the tub with candles. He also ran an extension cord from the next room, plugged in a radio, and left it on a nearby bench. Then he knocked the appliance off its perch and into the tub—but his wife caught the radio. She later brought charges after discovering he had searched the Web for "electrocution in a bathtub."

HORROR FACTOR: 8

Paralyzed, burned, and naked is a tough way to go.

GRIM FACTS

- A German study found that 75 percent of electrocutions in the bathtub were caused by hair dryers. Other electrical appliance culprits included table lamps and telephones.

- Fish swimming in a lake struck by lightning survive because the current travels across the surface rather than electrifying the entire body of water. A fish that happens to be nipping a fly off the surface when lightning strikes does run the risk of electrocution.

A Window to the Soul

Toronto lawyer Gary Hoy plunged to his death when he smashed through a glass wall on the twenty-fourth floor of the Toronto Dominion Bank Tower on July 9, 1993. Hoy was demonstrating the strength of the glass in the firm's offices—he believed the glass was unbreakable—to a group of visiting law students.

HEART ATTACK

Instances of heart attack increase with age, but anyone with enough gunk junking up their arteries is susceptible. If you're old enough to eat doughnuts and operate a remote control, congrats—you've reached the legal minimum age for a fatal attack.

HOW IT KILLS

When someone keels over dead from a heart attack, the murmurings at the funeral home are all about what "gave" the victim the heart attack, whether it's shoveling snow, picking up a fat grandchild, or engaging in some kind of indoor acrobatics unsuitable for an older someone (death by passion, they sometimes call it).

But the primary cause of heart attacks is coronary artery disease, also known as carotid artery disease. CAD is caused by a long-term buildup of cholesterol and plaque in arterial passageways. Eventually, the clogged arteries harden and can't pass enough oxygen-rich blood through to the heart, causing pain in the chest known as angina. If you were to drop while clearing snow or sweating off a few pounds with a friend, it probably was a long time in the making. So don't beat yourself up for doing whatever it was that made your heart race. You really had it coming.

Sometimes chunks of plaque break off and cause a clot that blocks the artery entirely and cuts off blood supply to the heart. Blockage from a clot can kill heart muscle. An incapacitated muscle disrupts the normal and patterned pumping of the heart. It's not hard to follow the quick chain reaction from there. No pumping heart means no blood; no blood, no oxygen; no oxygen, no life.

Early symptoms of a heart attack include pressure in your chest, abdominal discomfort, and shortness of breath. The squeeze on your chest increases so that you feel as if you are trapped under the weight of a grand piano. Panic will quickly set in, and you'll break out in a sweat as pain moves down your arm, shoulder, neck, and/or back. The throbbing may even radiate to unlikely places in your body such as the jaw and teeth. Before passing out, you might experience dizziness, nausea, and vomiting.

Sudden cardiac death, aka sudden arrest, is caused by CAD 90 percent of the time, but it's also possible with no history of heart disease. Adrenaline from intense physical activity can also stop the heart, as can the use or abuse of drugs, be they illegal or prescribed.

KNOWN BY SCIENCE AS:

Myocardial infarction

MEDICAL CAUSE OF DEATH

Loss of heart function; cardiac arrest

TIME TO KILL

From minutes to months

HIGHEST RISK

Obese males over the age of sixty-five with family history of heart disease

LETHALITY

Medium—in fact, dead center. Half the people who have heart attacks in the United States do not survive them.

KILLS PER ANNUM

Approximately 500,000 in the United States, amounting to about one in five of all deaths.

HISTORIC DEATH TOLL

Worldwide, over 1.8 billion in the twentieth century. The death rate from heart disease has declined dramatically in the past fifty years thanks in large part to improvements in health care and in emergency treatments. Still, heart disease remains the leading cause of death in the United States. The increasing prevalence of obesity, exacerbated by physical inactivity and smoking, is slowing an otherwise positive trend.

NOTABLE VICTIM

Those urged to exercise against their will in the name of good heart health love to remind skinny people about Jim Fixx. The forerunner of America's fitness movement died of a heart attack in 1984, at the age of 52, shortly after a routine run.

HORROR FACTOR: 4

The last few minutes won't be much fun, but it can be a quick way to go.

GRIM FACTS

- Heart attacks don't always look like they do in Hollywood scenes, where the victim very suddenly grabs his chest, knocks over a vase, and falls to the floor. Sometimes they strike with no symptoms at all—even while the victim is at rest.

- Women exhibit some symptoms of a heart attack that men do not. Abdominal pain, heartburn, clammy skin, dizziness, and fatigue are all fair game for the fairer sex.

- The World Health Organization (WHO) expects chronic conditions like heart disease and stroke to overtake infectious diseases as the leading cause of death by the year 2030.

HELMINTH PARASITES

Are you the host with the most?

HOW IT KILLS

Few medical maladies are as sickening to imagine as parasitic worms entering through your mouth, producing larvae in your intestines, and living—sometimes for years—on the nutrients sucked from your body's organs. This is not a special-effects-laden science-fiction movie, mind you, but the real world. Parasitic infections afflict billions every year, all across the globe.

These helminth or worm parasites (as opposed to ectoparasites like mosquitoes and lice or single-celled protozoa parasites) typically enter through the mouth when fecal-infected food or water is ingested. They can also burrow right through the skin while a person is swimming or bathing in unsanitary water. Depending on a particular parasite's life cycle, they can find their way into a host in the larvae stage and hatch inside the body's warm environment, or crawl in as developed worms and lay new eggs. In the adult stage, a few worm types grow to be twenty inches long, and thick as a pencil. Imagine a lo mein noodle, but with a mouth. Now

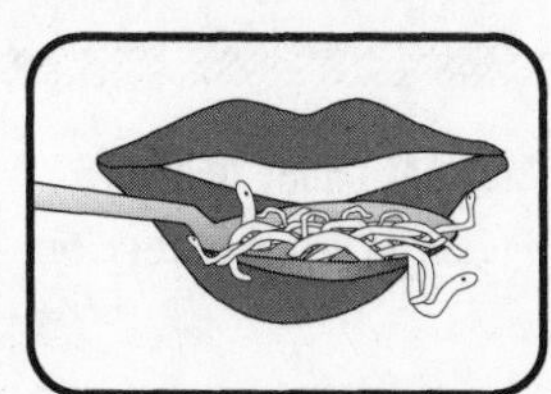

imagine a whole bowl of lo mein, because you never get just one parasitic worm.

Worms come in many shapes and sizes, and each type stands to infest and infect the host in a different way. Most will burrow through the intestinal wall or travel through the bloodstream to invade other organs. Untreated, several types of parasitic infections can kill. Among them:

Hookworm infection: Hookworms, which cling to walls of the digestive tract with a disgusting sharp-toothed mouth, can cause anemia. In severe cases, notably in children, anemia will cause heart failure.

Toxocariasis: The feces of dogs and cats can contain eggs of the Toxocara roundworm, which hatches in the intestines and usually heads for the liver or lungs. Toxocariasis can cause spleen infection or pneumonia. It can also disrupt vision if the migrating larvae infect the eye.

Ascariasis: The Ascaris worm, also known as the Giant Intestinal Roundworm, grows so big that it blocks ducts and tracts of the digestive system. Tangled clumps of worms larger than a softball can cause intestinal blockage unless the living clusters are vomited or passed through the stool. Migrating worms can also cause deadly conditions such as pneumonia and gangrene. This roundworm currently occurs in over one billion people worldwide.

KNOWN BY SCIENCE AS:
Infection by helminth parasite

MEDICAL CAUSE OF DEATH
Heart failure; infection; pneumonia

TIME TO KILL

A parasitic infection may not become life-threatening—or even be detected—for several years.

HIGHEST RISK

People in developing countries where sanitary water and hygiene education are lacking. Children—especially thumbsuckers—in these nations are at notably high risk.

LETHALITY

Medium. Survival depends on early detection and treatment.

KILLS PER ANNUM

Varies by parasite. In a year with 12 million acute Ascaris infections worldwide, there will be approximately 10,000 related deaths.

HISTORIC DEATH TOLL

Unknown, though fewer people are dying as nations acquire the resources to improve sanitary conditions.

HORROR FACTOR: 9

In some cases of Ascariasis, thick worms have been seen emerging from the throat and/or nasal passage.

GRIM FACTS

- You can get trichinosis, which is an infection by the Trichinella roundworm, from eating raw pork. The worms may travel all the way to your heart but they rarely kill.

- Tapeworms can grow up to thirty feet in length, and are sometimes felt moving out through the anus.

- New research suggests that some helminths actually support the immune system and may be helpful in fighting conditions like irritable bowel syndrome. Researchers have had human subjects deliberately swallow thousands of microscopic eggs.

Study Hell

Twenty-five-year-old Renée Hartevelt, a student at the Sorbonne, accepted an invitation to dinner from fellow student Issei Sagawa on the night of June 11, 1981. During the evening's conversation, Sagawa shot Hartevelt in the neck with a rifle, killing her. He then cut her up and ate her over the next two days. He was deported by the French back to his native Japan, but was released from jail a little over a year later.

HEMLOCK

Storied throughout history for its role in killing the Greek philosopher Socrates, hemlock is an herbal poison that comes in two lethal flavors: paralyzing and convulsing.

HOW IT KILLS

Hemlock grows wild throughout North America and Europe, reaching an average height of about seven feet. The seductive element of hemlock for humans is that its leaves resemble parsley while its taproots look like carrots or parsnips. Hence, those foraging for fresh vegetables may find a fresh way to die if they're not careful.

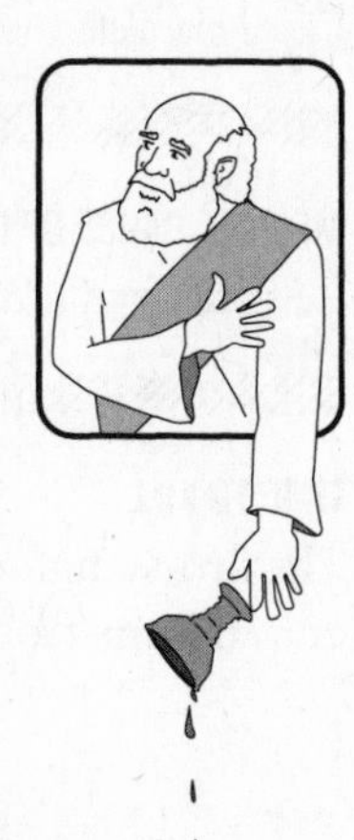

There are two types of hemlock: poison hemlock and water hemlock. The alkaloid toxins in poison hemlock have an effect on your body similar to nicotine. After you eat it, the poison goes right to work on your central nervous system. You'll have a stomachache and a headache, followed by unsteadiness and the shakes. Your heart rate will increase and you'll sweat and salivate uncontrollably. Then, in an interesting reversal—at least from a medical point of view—your heart will slow way down, and you will become paralyzed in your lower limbs. This paralysis will work its way up your body, gradually locking up your abdomen and chest. Right about now, you'll hope to go into a coma, which is the next stage of the poisoning as it completely overtakes your nervous system. While you're in a coma and your body is paralyzed, your lungs will be unable to suck in air, and you'll pass on to a garden where the vegetables aren't so poisonous.

Water hemlock contains a neurotoxin called cicutoxin, which is a strong convulsant. Like poison hemlock, it will make you nauseous and woozy. You'll also get the sweats, but water hemlock packs an extra punch: you'll suffer severe convulsions and then seizures. But these seizures are as bad as any you can ever get. Your brain will have one long, uninterrupted seizure that doesn't stop, known as status epilepticus, which will kill you as it frazzles your brain or sends you into a coma.

KNOWN BY SCIENCE AS:

The scientific name for poison hemlock is *Conium maculatum*. It is derived from the Greek word *konas* (to whirl), as one of the main symptoms of hemlock poisoning is loss of balance. Water hemlock is *Cicuta maculata*.

MEDICAL CAUSE OF DEATH

Respiratory failure; asphyxiation; status epilepticus

TIME TO KILL

The first symptoms will show up after fifteen minutes; death can occur anytime from two to twenty-four hours later.

HIGHEST RISK

People who mistake hemlock for wild carrots; Asians who make herbal home remedies; philosophers who upset the status quo

LETHALITY

Medium, around 30 percent. Many people vomit up hemlock before it can do much damage. For severe cases, there is no antidote.

KILLS PER ANNUM

Estimated one dozen

HISTORIC DEATH TOLL

Not known but likely in the thousands. These are mostly intentional poisonings—including capital punishment—and children eating the plants.

NOTABLE VICTIM

The Greek philosopher Socrates was killed by hemlock. In 399 B.C. he was found guilty by a Greek court of being a bad influence on local children and of not worshipping the ancient gods. His death sentence required him to drink a goblet of hemlock.

HORROR FACTOR: 5

You're unlikely to know what you've gotten yourself into until the paralysis or convulsing starts. Up until then, it feels like bad food poisoning.

GRIM FACTS

- Water hemlock is the most poisonous plant indigenous to North America.

- The hemlock tree, a conifer, is unrelated to either the poison hemlock or the water hemlock plants.

- Asian communities around the world are in increasing danger of hemlock poisoning as the plant has been found in packets of herbal supplements that are taken as vitamins or steeped to make tea.

- The leading right-to-die advocacy group in the United States was originally called "The Hemlock Society."

HEPATITIS

A veritable alphabet soup of deadly viruses, hepatitis has become a prominent killer throughout the world, putting more people on the waiting list for liver transplants than any other disease.

HOW IT KILLS

Hepatitis is an inflammation of the liver, typically caused by one of five viruses, helpfully labeled as the A, B, C, D, and E viruses.

Each variation of hepatitis has its own idiosyncrasies, depending on which virus you're infected with. Hepatitis B and C are two of the most virulent, spread via bodily fluids or skin puncturing (such as a hypodermic or tattoo needle). Hep A can come from contaminated food. Some forms, such as D, only occur once you've contracted one of the others.

After you contract hepatitis, it takes about three days for the virus to go to work on your liver. It begins replicating inside your liver cells, causing swelling and infection that will affect your liver's ability to break down dead blood cells in your body. When the first symptoms show up, you're likely to mistake the feeling for the flu, with the requisite aches and pains, general fatigue, nausea, and headache that go with it. Then you'll start having abdominal pains, accompanied by vomiting and diarrhea. As the disease progresses, you'll become even more tired and will suffer a loss of appetite. The buildup of dead blood cells will eventually cause your urine to darken. Then jaundice will set in—turning your skin, and the whites of your eyes, yellow.

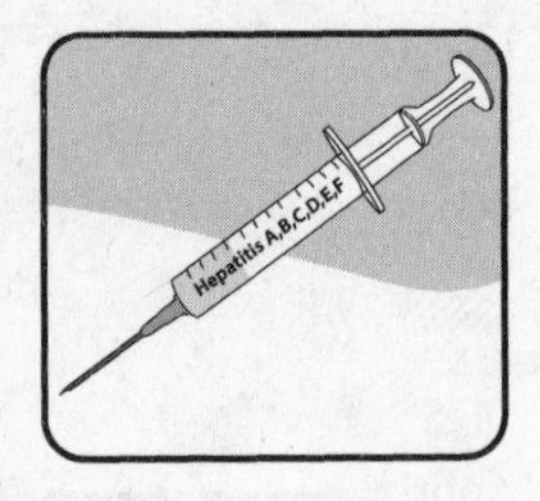

The constant assault of infection and inflammation in your liver will cause it to harden and then scar, a condition known as cirrhosis. Cirrhosis is basically the point at which your liver ceases to be able to repair itself or to function properly. This will result in your liver being susceptible to cancer, or failing completely. Once your liver fails, you need a new one—from someone else—or you're dead.

KNOWN BY SCIENCE AS:

The hepatitis B virus, or HBV, is a hepadnavirus, part of the family of viruses that infect human livers. Hepatitis C, or HCV, was originally called "non-A, non-B hepatitis" after its discovery in 1989.

MEDICAL CAUSE OF DEATH

Liver failure; liver cancer

TIME TO KILL

Years. Symptoms may not show for several months, and even then, it takes a while for the virus to dismantle the delicate machinery of the liver to the point where your liver can't do its job.

HIGHEST RISK

Drug users sharing needles; careless sex partners; hospital workers; patients on dialysis; tattoo fanatics; winos

LETHALITY

Low. Death occurs primarily in those with chronic or long-term hepatitis, and only in about 25 percent of those cases.

KILLS PER ANNUM

Approximately 1.5 million

HISTORIC DEATH TOLL

Millions. And the number of deaths increases every year.

NOTABLE VICTIM

Ken Kesey, the author of *One Flew Over the Cuckoo's Nest* and founder of the Merry Pranksters, died of liver failure related to Hepatitis C on November 10, 2001.

HORROR FACTOR: 3

Primarily a long-term disease, there isn't a lot of horror associated with hepatitis. However, people with Hep C are almost guaranteed to be waiting for either a transplant or an undertaker at some point during the infection, so we'll give Hep C a 3.

GRIM FACTS

- Outside of inflammation caused by viruses, hepatitis can also occur as a result of sustained alcohol abuse.

- There are vaccines to prevent Hepatitis A and B, but no vaccine exists for the Hepatitis C. And the Number 1 reason for liver transplants in the United States is Hep C.

- Liver cancer, often associated with the various forms of hepatitis, is the fourth leading cause of cancer-related deaths worldwide.

- Half a billion people worldwide are currently infected with Hepatitis B or C, ten times the number infected with HIV/AIDS. Researchers believe that most of those infected are not aware they have the disease.

HIV/AIDS

Once upon a time—not so very long ago—the most dangerous thing about sex was that you'd have to get penicillin shots for syphilis. With the introduction of AIDS into the equation during the 1980s, sex is now almost as lethal as Russian roulette with a fully loaded gun.

HOW IT KILLS

AIDS is caused by the human immunodeficiency virus, known as HIV. Technically, AIDS is a syndrome that is indicated by a group of symptoms or diseases that occur in the last stage of HIV infection. These conditions are caused by the overwhelming breakdown of your immune system, which allows all sorts of diseases to invade your body.

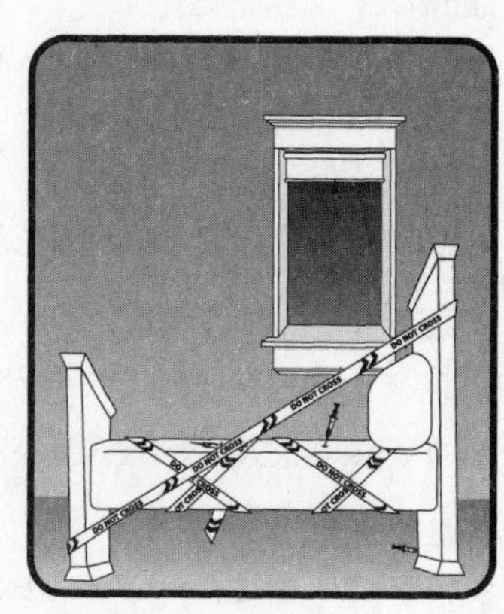

Let's start at the beginning. HIV is spread between two people via the exchange of body fluids, primarily semen and blood. Thus it can be passed on via sex, previously used hypodermic needles, and even infected blood from transfusions. Breast milk is also a source, which accounts for a huge number of afflicted children worldwide, though the virus can also be transmitted from mother to child before or during birth.

After you are infected, the virus starts destroying your white blood cells. Specifically, it attacks CD4+ T cells, which are key elements in maintaining the strength of your immune system. Initially, however, you'll experience no specific symptoms other than a general flulike feeling. But once your immune system starts breaking down in earnest, you're fair game for any disease that might want to come your way. That includes opportunistic diseases like pneumonia and cancer, which will slowly but surely prey on your weakened body.

You'll start displaying symptoms associated with the individual opportunistic diseases, ranging from difficulty in breathing to gastrointestinal pain. By this time, you will have full-blown AIDS, which finally has some symptoms to call its own. These include severe weight loss and anemia, diarrhea, weakness, fatigue, and swollen lymph nodes.

Ultimately what kills you depends on the disease or infection that hits you once your body simply can't take any more. In effect, then, AIDS itself doesn't kill you. It is something far more common and for many, this is pneumonia.

KNOWN BY SCIENCE AS:

AIDS, or acquired immune deficiency syndrome (also called acquired immunodeficiency syndrome), is caused by the human immunodeficiency virus (HIV).

MEDICAL CAUSE OF DEATH

Depends on the type of disease in the final infection

TIME TO KILL

A year or more, sometimes longer than a decade

HIGHEST RISK

Promiscuous gay men; prostitutes and their customers; drug users sharing needles; babies born in sub-Saharan Africa; patients getting blood transfusions from substandard supplies

LETHALITY

Medium. There is no known cure, but the symptoms can be moderated by medication. Some of the opportunistic diseases associated with AIDS can be cured, but the disease will never go away.

KILLS PER ANNUM

Three million annually, of which roughly a sixth are children under the age of fifteen

HISTORIC DEATH TOLL

Forty million and growing, with no end in sight

NOTABLE VICTIMS

Movie icon and man's man Rock Hudson was the first publicized celebrity AIDS death when he died on October 2, 1985. He was followed on November 24, 1991, by rock icon and singer for the band Queen, Freddie Mercury, who died from pneumonia as a complication of AIDS. Both men contracted the virus from gay sex.

HORROR FACTOR: 4

Because AIDS is survivable, there is hope on the part of victims that they can outlive the disease.

GRIM FACTS

- HIV is believed to have originated in West African chimpanzees. Hunters became infected with the virus when they hunted these apes for food and were exposed to their blood.

- The first citation of AIDS as a medical condition was June 5, 1981, when the Centers for Disease Control (CDC) recorded a group of unusual pneumonia cases in five gay men living in Los Angeles. HIV has been traced in humans as far back as 1959.

- AIDS is not transmitted in the air, nor can it be spread by mosquitoes (HIV is digested and destroyed inside the insect).

- AIDS is so prevalent in third-world countries that it is on track to become one of the top three killers in the world by 2030, after heart disease and stroke.

HOLE IN THE HEAD

Yeah, you need a lot of things like you need a hole in the head. And a hole in the head is one of them.

HOW IT KILLS

You have inadvertently walked onto the local archery range during tryouts for the freshman team, and a wayward arrow finds its way to your skull. This arrow actually enters your skull, which is a big deal since it takes a lot of concentrated force to penetrate the thickness of your skull and put a hole in your head.

Pointed objects, primarily those with widths of less than an inch, will pierce your skull, and your brain, without taking off your entire head. They must be traveling very fast—or your skull must be. Arrows, bullets, spears, metal projectiles, and stationary objects, like spikes and rebar onto which your head impales itself, all fall into this category.

Your skull is composed of eight different bones that fit together like a spherical jigsaw puzzle. They are the frontal, occipital (back), a pair of parietals (top sides), a pair of temporals (lower sides), and the sphenoid and ethmoid, which connect the rest of the skull to the face bones. The thickest section of your skull, around the back and sides, is slightly less than half an inch thick.

Once the skull is pierced by this arrow—which is traveling at about 175 miles per hour—bone fragments push forward into your meninges, ripping the Saran Wrap–like covering that keeps your brain free from bacteria and viruses. The tip of the arrow, moving forward, tears into your cerebrum, which is the outer section of your brain responsible for thinking. The cerebrum's gelatinlike consistency allows the arrow to pass through with little resistance. The arrow severs millions upon millions of connections between your neurons. Your brain has been making these connections since you were born.

As it slices through one hemisphere of your brain, the arrow ruptures and rips blood vessels. Blood—and there is a great deal of it in your brain—starts emptying out of your head through the entrance wound made by the arrow. The arrow is still moving and pierces the corpus callosum, which connects the two halves of your brain. The arrow flies through your other hemisphere and forces its way out just like it came in—by punching a hole out of your skull.

If the arrow has gone directly from one side of your head to the other, you will lose speech and some cognitive functions. If it goes from front to back, you will go blind. If it goes slightly askew and hits your cerebellum or brain stem, you will be paralyzed. All of this is moot, of course, because an object that rips through your entire head—leaving two holes for blood to spill out of—is going to deprive you of the oxygenated blood your brain needs to function. And then it's lights out.

KNOWN BY SCIENCE AS:

Traumatic brain injury, or TBI, caused by a projectile or stationary object

MEDICAL CAUSE OF DEATH
Brain damage

TIME TO KILL
Instantly, if you're fortunate; otherwise, a few minutes

HIGHEST RISK
Archers; sniper targets; gang members; construction workers

LETHALITY
Medium. It all depends on what parts of the brain get hit by the object traveling through. The deeper in the brain the damage occurs—particularly around the brain stem—the higher the odds of death.

KILLS PER ANNUM
Of the 50,000 people who die from TBI in America each year, an estimated 25 percent of those involve puncturing the skull (as opposed to simply smashing it). With 12,000 in the United States alone, and no planetary records to be found, estimates are about 75,000 worldwide.

HISTORIC DEATH TOLL
Millions. Ancient skulls have been found with holes punched in them, indicating murder, accidents, and even religious ceremonies that allowed headaches to escape.

NOTABLE VICTIM

Famed murderer and bandit Jesse James was hunted unsuccessfully by law enforcement for almost twenty years. Jesse's good friend Robert Ford brought that all to a halt with a bullet to the back of Jesse's head—administered on April 3, 1882, while James was cleaning off a dusty picture.

HORROR FACTOR: 1

Once you get TBI from a hole in the head, it's pretty much over in no time.

GRIM FACTS

- Pre-Colombian cultures in South America punched holes in people's skulls with metal blades and sharpened stones for reasons that aren't quite clear. The best guesses are that it was done to relieve headaches or to release evil spirits, and it's a good guess that many of the "patients" died pretty soon thereafter.

- Numerous people have survived having their skulls and brains penetrated. On June 6, 2008, George Chandler of Kansas had a 2.5-inch nail accidentally driven into his head by a nail gun. A local doctor removed it with a clawhammer and Chandler suffered no ill effects.

- On September 13, 1848, Phineas Gage, a railroad worker, had a three-foot spike blasted up through his cheek and out the top of his head. He survived and was able to work again, but the spike destroyed his personality and the only living creatures he could stand to be around were horses. Gage remains the definitive study of TBI from a projectile.

HOUSE FIRE

The prospect of your home going up in flames is as petrifying today as it was when you were five. Every animal on the planet fears fire, including the hulking firefighters who carry kids, grannies, and dogs from burning buildings. Evolution has trained us to fear and respect the flame, and that goes a long way toward keeping people safe. But not far enough.

HOW IT KILLS

Most house fires start in the kitchen, and you don't have to be cooking up anything fancy to flambé the cupboards and curtains. Even among the cautious and fire-conscious—that is, people who don't leave clothes near candles or gas cans near a fireplace—a residential fire can ignite from faulty wiring, a malfunctioning appliance, or even a lint-filled dryer. Your vices can get you, too; smoking cigarettes is the leading cause of all fires, and alcohol is a huge contributor to fire-related deaths. So if your partner has had a few pops and

wants to enjoy a cigarette before rolling over, ask her to sleep out in the driveway.

Waking to a room thick with smoke, you would do well to remember the classic survival technique Fireman Bill taught your first-grade class: stop, drop, and roll. In case you don't remember, you want to "drop" because smoke rises—and the inhalation of smoke, or of the toxic fumes rising from your smoldering possessions, is likely to asphyxiate you long before the fire comes close enough to burn your skin. Even if the flames are near, you are less likely to burn than to suffer the fatal consequences of inhaling superheated air, which scalds and scars the fragile airways of the respiratory system where oxygen is exchanged with the bloodstream.

At the same time—and this is very bad timing—panic causes your heart to race and your breathing to become shallow; the body's demand for oxygen is at an all-time high while supply is at an all-time low. Unless you can make your way to cool, clean air, you'll die a choking and breathless death within minutes right there on the ol' shag rug.

KNOWN BY SCIENCE AS:

Residential fire emergency

MEDICAL CAUSE OF DEATH

Asphyxiation

TIME TO KILL

In heavy smoke and hot flames, two to five minutes

HIGHEST RISK

Children under four and elderly people living in homes without smoke detectors; having an alcoholic in the house—especially one who smokes in bed—sends risk through the burning roof.

LETHALITY

Medium. For every hundred people injured in a house fire, nearly twenty die.

KILLS PER ANNUM

Between 2,500 and 3,000 in the United States

HISTORIC DEATH TOLL

About 450,000 in the twentieth century. For fifty years prior to the mid-1980s, residential fires killed an average of 5,000 people every year in the United States. The advent of fire safety education and the ubiquity of smoke alarms (now in 90 percent of all homes) has halved the statistic.

NOTABLE VICTIM

Jack Cassidy, the debonair character actor and father of teen icons Shaun and David, died in December 1976 when his West Hollywood apartment went up in flames. Reports at the time said Cassidy fell asleep on his couch—with a cigarette in hand—after a late-night Christmas party.

HORROR FACTOR: 9

It's fire. Fire!

GRIM FACTS

- On average in the United States, someone dies in a house fire every 162 minutes.

- Alcohol not only increases your chances of starting a fire by impairing your judgment, but decreases your chances of waking up and escaping it. More than 40 percent of all adults killed in house fires are found to have been under the influence.

- Fire is the third most fatal home injury, following falls and poisoning.

HUNGER STRIKE

Every system, every organ, every cell in the human anatomy depends on food. By ingesting nutrients, we provide the body with energy. It's not just a matter of having your Rice Krispies and OJ so you feel peppy in the morning. The energy provided by food enables all of the organic functions—metabolism, growth, cell reproduction—that differentiate a live person from a slab of meat. Without food, you're not much more than that.

HOW IT KILLS

Given how enmeshed food is with human life, it's remarkable how long a healthy, well-nourished person will live if suddenly cut off from all nutrition. What with our three meals a day, and with our overindulgence of the slightest hunger pang, we can survive more than two months without consuming any food for energy. But oh what a dismal two months it would be.

The mechanics of death by starvation are as varied as your body's systems. That is to say, every part of the body responsible for keeping you alive will eventually fail if you are starved. In the earliest stage, your body starts tapping the carbohydrates tucked away in your muscles and your liver. This leads to a quick loss of weight at the expense of muscle mass and liver function. The carbs burn up quickly, so your

body turns next to the energy stored in fat. As anyone who's tried to lose weight by more healthy means knows, burning fat takes a long time—which, in a starvation scenario, is good news since it means the energy lasts longer and leaves the pounds on your bones for a longer period of time.

You will likely experience some stomach cramps in the first few days without food. Don't worry, they'll pass—it's just that your digestive system is shrinking. Around the two-week mark, though, symptoms start to get really bad. You're dizzy and lethargic. Your neuromuscular system becomes sluggish and you begin to have difficulty standing up. As you lose the ability to control muscle movement, you can't sit down too well, either. By Day 15 or 20 you're getting extremely weak, and without a fuel source your body is having trouble staying warm. Erratic changes to your breathing and swallowing lead to uncontrolled bouts of the hiccups. And every hiccup hurts. You're not looking too good: bones are starting to protrude and your skin has gone pale (after all, you're anemic now). By the end of the first month, all sorts of body parts are shriveling and drying up: your hair is falling out, your tongue is like a piece of leather, and your testicles (or ovaries) are shrinking. This is not going well at all.

At about six weeks without food, you can't keep your eyeballs from rolling all around in your head. Someone tries to give you water, but you can't swallow.

Even if you're hydrated intravenously, your body can't hold on to the liquids—you may even die of dehydration from diarrhea. Like every other muscle in your body, your heart has shrunk and labors to pump within the low-blood-pressure environment. Your breathing is shallow. The cold from lack of fuel and reduced blood flow has reached hypothermic levels now and fluid is beginning to accumulate in your arms, legs, and abdomen just as it does in a dead body. In fact, by now you've taken on most of the distinguishing characteristics of a corpse, and within eight or twelve weeks—by means of heart failure, respiratory failure, or neurological shutdown—you are one.

KNOWN BY SCIENCE AS:
Starvation; acute undernutrition

MEDICAL CAUSE OF DEATH
Heart failure; respiratory failure; dehydration; brain death

TIME TO KILL
Eight to twelve weeks

HIGHEST RISK
Hunger strikers; severe anorexics; the millions of undernourished people worldwide who have limited access to food

LETHALITY
As high as it gets. There's no upside to starving.

KILLS PER ANNUM
A politically turbulent year can lead to dozens of hunger-strike fatalities. For those not intentionally starving themselves, the Food and Agriculture Organization of the United Nations estimates that more than 25,000 people die of starvation every day. That's 9,125,000 every year.

HISTORIC DEATH TOLL

Several dozen by hunger strike in the twentieth century (most of them political prisoners and dissidents)

NOTABLE VICTIM

Bobby Sands remains the Grand Poobah of hunger striking. He died after sixty-six days without food in a Northern Ireland prison—during which time he was also elected to Parliament. Of the twenty-two other Irish Republicans to participate in the famous 1981 Hunger Strike, another nine died alongside Sands.

HORROR FACTOR: 7

Pick a cause you really believe in, because this is one long and torturous way to make a point.

GRIM FACTS

- In "rolling" hunger strikes, one striker after the next is called on to starve himself in support of a cause. However, the plan is to pass along the fasting mantle before anyone dies.

- In April 2006, five students and six janitors at the University of Miami went on a hunger strike to protest university president Donna Shalala's position in a labor dispute. Maybe they'd all been studying irony: Shalala had served under Bill Clinton as the secretary for Health and Human Services.

- According to the World Health Organization (WHO), nearly one in three people on Earth dies or is disabled from poor nutrition or calorie deficiency.

JUMPING OFF A BRIDGE

One thing about jumping off a bridge . . . there's no way to rethink it once you're headed for the water.

HOW IT KILLS

Suspension bridges that rise high above rushing water have always attracted those who want to find quick death. The San Francisco Golden Gate Bridge, situated 220 feet above the turgid waters of the Golden Gate strait, has had more than 1,200 people jump to their deaths off of its majestic deck.

The thing about jumping off a bridge into water from a great height is that it's not much different than diving into concrete from a great height. You are hitting the water at a rate of over 80 miles per hour. Not to be overly technical about it, but at that speed, the water doesn't have much time to get out of your way. You hit with such force that the water acts like a solid when it is compressed by the weight of your body. You sink under only after you've come to an abrupt stop.

And as with all incidents involving massive trauma, this one isn't pretty. When your body hits the water, your rib cage will crack, and flopping limbs will break. Your spine will snap and vertebrae will shatter while your skull fractures. Most of the blood vessels where your body makes contact with the water will rupture. Your organs, which are not even close to being cushioned by your skin, will slam forward with the force of a NASCAR crash. They will rip apart from each other and then collapse into themselves, crushed like soft pieces of fruit. Blood then flows freely into your body cavities.

Plus, there's an added complication. Even if the trauma doesn't kill you immediately, your body will have suffered so much internal damage that you will either be unconscious or unable to swim. Either way, you'll drown. And that will kill you.

KNOWN BY SCIENCE AS:
High-velocity impact with water

MEDICAL CAUSE OF DEATH
Massive trauma; asphyxiation

TIME TO KILL
From the moment you jump till the time you hit the water: four seconds

HIGHEST RISK
Mentally and emotionally depressed people; those who purchase secondhand bungee-jumping gear; tourists who "want a better look" from the bridge

LETHALITY ☠☠☠
Almost 100 percent. If the bridge is high enough, you're going to hit with so much force that nothing short of a parachute is going to

save you. A nice feet-first dive has been known to prevent instant death.

KILLS PER ANNUM

Worldwide estimate, about 100. The Golden Gate Bridge alone has a jumper on average of once every two weeks.

HISTORIC DEATH TOLL

Several thousand. Bridges high enough to cross large bodies of water—and therefore attract jumpers—have been around just over a century. However, ancient rope bridges strung throughout the world's jungles most assuredly had their fair share of people who swan-dived into the river below.

NOTABLE VICTIM

While walking with her parents on December 22, 1997, two-year-old Gauri Govil slipped through a nine-and-a-half-inch gap in the railing of the Golden Gate Bridge. The little girl plunged nearly two hundred feet to her death.

HORROR FACTOR: 7

Once you jump, you're done—but you get four seconds to give some serious thought to how bad it might hurt.

GRIM FACTS

- The Golden Gate Bridge in San Francisco, the Price Edward Viaduct in Toronto, and the Aurora Bridge in Seattle are considered "suicide magnets," and have nearly 2,000 deaths among them.

- At over 1,000 feet, the Millau Viaduct in southern France is the world's highest traffic bridge. The first suicide occurred exactly one month after the December 14, 2004, opening of the bridge.

- Contrary to popular belief, you do not pass out when falling from a great height. Skydivers regularly free-fall thousands of feet. This means you don't pass out before you hit bottom; you're very much alive on impact.

Electrifying Performance

Les Harvey, guitarist for the band Stone The Crows, touched an ungrounded microphone at the Top Rank club in Swansea, Wales, on May 3, 1972. He was electrocuted and died onstage.

KNIFE WOUND

Knives are found in every culture, and in lots of people's backs. We're going to take a stab at explaining how these pieces of metal can cut the life out of you.

HOW IT KILLS

Knives are pieces of metal that have been shaped and sharpened into a blade. Many knives are designed as weapons, but unless you're the target of an assassin, you're most likely facing someone who's using the nearest readily available knife, and odds are that it will be a kitchen knife.

Knives come in a variety of lengths, but standard kitchen knives range from the 4.5-inch blade of a steak knife to the eight-inch blade of a knife used for cutting meat or chopping vegetables. Given the flimsy construction of most steak knives—the blades are apt to snap if thrust against bone—we'll focus on the long knives.

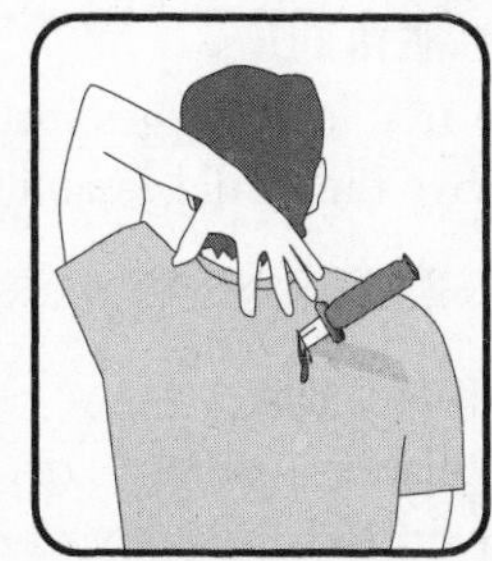

Your assailant is likely to attack by aiming for your chest or neck, which are natural targets for someone holding a blade aloft and thrusting it downward (long knives are deadliest when thrust with the force from your elbow). You will bleed immediately from any contact between the blade and your skin, which will be a slash or a puncture that cuts open capillaries and other blood vessels. While a single stab wound is unlikely to kill you, repeated stabbings will. The slicing of both veins and arteries will result in heavy blood loss, which may be enough to kill you.

You will also incur what are called "defensive wounds" as you ward off the attack with your hands and arms. This will increase blood loss, leading to shock, not to mention a great deal of pain.

The puncturing of a vital organ—stomach, intestines, liver, kidney, and even heart—can cause organ failure. Thanks to its length, a long knife can be driven most of the way through the chest cavity or abdomen when pushed in "up to the hilt," as it were. Yet this can only occur if the attacker attacks using an awkward grip that allows stabbing forward, upward, or from the side. Unless you're incapacitated, these stab wounds are likely to be the exception rather than the norm.

KNOWN BY SCIENCE AS:

Puncture wound

MEDICAL CAUSE OF DEATH

Organ failure; blood loss; shock

TIME TO KILL

Variable. It ranges from a few minutes to an hour or so. If you survive the initial assault, you might just be okay.

HIGHEST RISK

Relatives; ex-girl- or boyfriends; jealous lovers; sentries on night patrol; drunken bar brawlers; neighbors of knife-wielding maniacs

LETHALITY

Low. Stab wounds to most of the body, notably the extremities and even the face and head, will not result in death. The rib cage prevents direct assault on the most vulnerable organs, so the attacker has to be skilled enough, or lucky enough, to strike unprotected organs (liver, stomach, kidney, intestines) or sever major blood vessels.

KILLS PER ANNUM

Between 1,500 and 2,000 every year in the United States alone, with an estimated 20,000 more worldwide

HISTORIC DEATH TOLL

Millions. People have been stabbing each other to death from the moment they could sharpen a bit of stone or piece of metal.

NOTABLE VICTIM

Julius Caesar was stabbed on March 15, 44 BC by members of the Roman senate. The assassins stabbed Caesar nearly two dozen times before he died. He uttered his famous last words to his best friend Brutus, who was an accomplice in the murder: "*Et tu, Brute?*" (You too, Brutus?). That day, Brutus became the quintessential "backstabber."

HORROR FACTOR: 8

Getting stabbed once is bad, but it really takes a prolonged attack to kill you. The prolonged aspect is increasingly painful and the attendant blood spattering will freak out anyone.

GRIM FACTS

- Stabbing yourself to death with a knife in a ritual suicide in Japan is called *seppuku* (commonly referred to as hara-kiri).

- Stabbing is the most common form of murder in Britain, accounting for nearly a third of all homicides (guns are outlawed in Britain, but are still used in roughly 10 percent of all murders). Stabbing accounts for 13 percent of U.S. murders.

- If the knife is left sticking in your body after a single stab, there may not be much blood. But just because you're not bleeding on the outside does not mean that you're not gushing a lot on the inside.

- Other instruments used for stabbing—both intentional and accidental—include box cutters, razors, glass shards, nails, scissors, pens and pencils, drill bits, and metal spikes.

KOMODO DRAGON

Some reptiles are likeable little creatures you can keep at home as interesting pets. Others would like nothing better than to dine on your decaying flesh. The Komodo dragon falls neatly into the second group.

HOW IT KILLS

Death by Komodo dragon can be horrific and slow. This largest of all living lizards typically ambushes a victim by lunging first at the throat and then gorging the underbelly. Holding down its prey with powerful front legs, the Komodo takes about fifteen or twenty minutes to eat a goat-sized meal.

If the initial attack doesn't kill you, the virulent bacteria in a Komodo's mouth from even a nibble on the leg would render you lifeless within a few days. Once the septic bacteria does its deadly work, a pack of Komodos can smell your dead flesh from a few miles off and return to dine on your corpse. They'll also eat the bones, although after a meal they typically regurgitate parts they don't care for, like hair and teeth.

Fortunately for you, Komodos live only on a handful of islands in Indonesia; notably, Komodo Island. Tour guides on Komodo love to recount tales of unfortunate hikers or sunbathers who have disappeared in dragon territory, though their gory stories often lack real evidence. But it is true that Komodo dragons love carrion, and they've been known to dig up human remains from shallow graves.

KNOWN BY SCIENCE AS:

Varanus komodoensis; Ora; "land crocodile"

MEDICAL CAUSE OF DEATH

Bloodletting and/or suffocation; virulent bacteria-causing septicemia

TIME TO KILL

If you're not eaten and digested immediately, septicemia from a bite will kill you within two to three days. Then other Komodos will descend on the human buffet within a week.

HIGHEST RISK

The Komodo's only natural habitat is a few isolated islands in Indonesia. You either have to be a dumb tourist or an even dumber local to let your guard down in Komodo territory and thereby risk an attack.

LETHALITY ☠☠☠

High. Once you're attacked, the odds of dying are nearly 100 percent. In addition to evisceration, blood loss, trauma, and infection, the situation is not helped by the fact that the nearest hospitals are over three hours away by ferry. If treated in time, the septicemia from a minor Komodo bite can be cured with antibiotics. That's a very big "if."

KILLS PER ANNUM 

About one per year

HISTORIC DEATH TOLL

Medical record-keeping is a novelty in Indonesia, and there is no archive listing death by Komodo. However, a large percentage of missing-persons reports have been attributed to the dragon, and total deaths over the past century are estimated to be in the dozens.

NOTABLE VICTIM

In 1974, elderly Swiss tourist Rudolf von Reding Biberegg went for a hike in dragon territory. His friends found only his hat, his shoe, his camera, and a pool of blood.

HORROR FACTOR: 10 

At the top of the "Oh-God-Please-Help-Me" horror meter scale. Most people would rather be tortured by the guy in *Saw* than be eaten by a Komodo dragon.

GRIM FACTS

- Komodo dragons eat only about twelve meals a year; even less if their meals include big game like horse, deer, and water buffalo.

- The Komodo's evil image is enhanced by its long, forked, yellow tongue.

- The Western world didn't even know the dragons existed until 1910.

Liquid Diet

Jennifer Strange participated in a "Hold Your Wee for a Wii" radio contest in Sacramento, California, on January 12, 2007. The object was to drink as much water as possible without urinating. After drinking nearly two gallons of water, Strange went home to her three children. She died that evening from hyponatremia or "water intoxication," a condition where too much water thins out the sodium levels in the blood. Making it all the more dreadful was the fact that Strange wasn't even the winner of the contest.

LEAD POISONING

Lead made for some nice-looking wall paint, but it makes a lousy meal.

The U.S government banned use of lead paint in 1978, but that still leaves city upon city of older homes painted wall to wall with a known poison. Lead typically makes its way into the body when people put something in their mouth that's covered with lead dust. While not a huge liability for adults, small children may be on all fours near dusty windowsills and on dusty floors, representing the greatest risk. It's true that children may also be exposed by regularly snacking on crispy wafers of flaked paint, but that risk is generally overstated. Just when we started to think it was a poison of the past, however, lead started showing up in painted toys.

Let's imagine you're a lot younger than you are now, and you've been savoring lead chip hors d'oeuvres. Even at low levels of exposure, elevated lead levels in your blood can lead to toxic reactions in your kidneys and brain. Other tissues may be poisoned as concentrations of lead increase in blood and bone. The main problem for your body is that lead disrupts the production of hemoglobin, the protein that carries oxygen to red blood cells. Early symptoms of lead poisoning include sluggishness, vomiting, and a sickly skin pallor from anemia.

For children, the greatest risk is to brain development. Kids with lead poisoning are known to have developmental issues with speech, behavior, physical growth, and muscle coordination. In the most severe cases, nerve and blood disorders brought on by lead poisoning lead to seizures, loss of consciousness, and death.

KNOWN BY SCIENCE AS:
Acute lead toxicity; plumbism; painter's colic

MEDICAL CAUSE OF DEATH
Encephalopathy, a disease of the brain that alters brain function or structure

TIME TO KILL
Three weeks after significant symptoms arise

HIGHEST RISK
Crawling children in homes built before 1978, which are undergoing construction

LETHALITY
Low. Eating lead off the walls is nothing like "eating" a lead bullet.

KILLS PER ANNUM
Not even one per per year in the United States

HISTORIC TOLL

Difficult to measure, since it's hard to pin deaths specifically on lead intake, but estimated in the thousands. The removal of lead from gasoline, which took twenty-three years (from 1972 to 1995), led to a fourfold reduction in pediatric blood lead levels.

NOTABLE VICTIM

A two-year-old New Hampshire girl died of lead poisoning in 2000, the first U.S. pediatric death of its kind in ten years. An autopsy revealed the poisoning caused cerebral edema. The girl had moved into a 1920s-era home in New Hampshire with her Sudanese refugee family.

HORROR FACTOR: 5 

Really a 2, but anything with kids ups the ante

GRIM FACTS

- Five of every seven homes in the United States were built before the ban on lead paint was enacted.
- Most reported health problems result from creating lead dust, fumes, and chips while disturbing painted surfaces. Left alone, lead paint is generally not a threat.
- Lead also makes its way into the home through soil and through lead water pipes, which can release particles into tap water.
- Though children first tend to be exposed to lead around the age of two, the resultant learning disabilities are seldom recognized before they enter school at age five.

LIGHTNING ON THE GOLF COURSE

If you've ever watched someone else play eighteen long, long holes of golf, you've probably wished for death to whisk you away. However, the Grim Reaper is seldom so merciful. Nor does he have the courtesy to yell "Fore!" in advance of sending 300,000 volts of electricity down onto the fairway.

HOW IT KILLS

Playing eighteen holes of golf may seem like an innocent and peaceful way to spend an afternoon, but the sport has claimed lives in any number of ways. With golfer demographics skewing toward the elderly, many grandpappies and wealthy old white guys have found themselves in a bunker following a heart attack or a stroke on the course. Carts crash, players get brained by flying golf balls, and the shafts of golf clubs broken in anger have pierced this or that vital organ. There have even been accounts of alligators emerging from water hazards in Florida and making off with a mouthful of plaid and knickers.

But every now and then a hapless golfer is zapped by a bolt of lightning. Given all the metal clubs and hilltop exposures, it's surprising it doesn't happen more often.

Lightning is the Tiger Woods of electrical injuries. While it's tempting to compare a lightning strike to electrocution by a high-

voltage line, lightning is in a league of its own. A direct strike may not last as long as the jolt you get before a circuit blows, but in those few milliseconds the contact voltage reaches over 300 kilovolts—about 100 times the surge needed to cook meat in your microwave.

You heard thunder as you rounded the last dog leg, but just couldn't help swinging that lightning rod as you went for the birdie. Now a bolt has hit your Big Bertha club right on the head—and since she's in your hands, the current seeks a route right through you from the club head to the ground. First, you're thrown the distance of a nice chip and your shoes are blown off. At the point of contact, your hands are deeply burned and the melted rubber handle fuses with your melted skin. At the point of exit, your feet are scalded. But your body can't register local pain right now because the jolt has paralyzed you.

If you're like most lightning victims, the strike causes a giant fibrillation that immediately leaves you in cardiopulmonary arrest, meaning simply that your heart stops. Paralysis of the central nervous system also stops your breathing cold, and without the aid of artificial respiration you'll die from lack of oxygen. If the incapacitation of your circulatory system or respiratory system hasn't killed you, you may still suffer severe, fatal bleeding in and around your brain. At this point, someone from your foursome should wave to the golfers behind you. It's best they play through.

KNOWN BY SCIENCE AS:

Electrocution, plus, a good walk spoiled—like, *really* spoiled

MEDICAL CAUSES OF DEATH

Asystole or ventricular fibrillation; severe cardiac dysfunction; central nervous system failure; hypoxia; cerebral edema

TIME TO KILL

With a direct hit leading to cardiopulmonary failure, instantaneous

HIGHEST RISK

Floridian golfers playing in July, especially in the afternoon. The Sunshine State has twice the lightning casualties of any other state.

LETHALITY

Medium. Unprotected humans stand a one in five chance of surviving a lightning strike, though a direct strike—from the cloud to your Callaway driver, as opposed to a charge that runs through the ground, or "side flash" lightning that jumps from an object nearby—is more likely to be fatal.

KILLS PER ANNUM

About eighty people die from lightning-associated deaths every year. Specific stats for golfers have not been compiled, but according to *The New York Times*, 24.5 million golfers played nearly 500 million rounds in 1993, and only one golfer died from a lightning strike.

HISTORIC DEATH TOLL

Several hundred. Deaths from lightning have decreased by about 30 percent in the past four decades, thanks to better weather forecasting and improvements in health care.

NOTABLE VICTIM

Chef Jess Roybal and his girlfriend, Pamela Jaffe, were struck by lightning on July 30, 1997, while playing golf on the John F. Kennedy Golf Course in Denver, Colorado. Roybal was killed instantly, and Jaffe went into a coma. Jaffe's father, a lawyer, sued the city for negligence and intentional infliction of emotional distress. He lost.

HORROR FACTOR: 3 

If lightning scares you that badly, stay in the clubhouse.

GRIM FACTS

- Men are five times more likely to play golf than women—and four times more likely to be struck by lightning.

- Golfers who figure they'll be safe in the golf cart, since it has rubber tires, may be in for a shock. The metal roof and sides of an automobile are what offer some protection in a lightning storm—so don't expect much from your roofless, sideless cart.

LIONS, TIGERS, AND BEARS

The interesting thing about lions, and tigers, and bears is that they don't live anywhere near each other except in zoos. If you, however, cross any of their paths outside of a zoo, you are likely to be part of their digestive tracts in less time than it takes for us to describe it.

HOW IT KILLS

Lions, and tigers, and bears are the apex predators on their respective continents, meaning they are at the top of nature's food chain and unlikely to become another animal's prey. Despite the fears of Dorothy and the Scarecrow, you are not likely to meet any more than one of them at the same time, as lions are located in Africa, tigers are found in southern Asia, and bears live in North America, northern Europe, and in some regions of Asia. Your personal encounter with them, however, is likely to end up similarly dreadful no matter where you happen to meet them.

Even though bears are not related to lions and tigers, all three share the ingredients that make them the most efficient killers on dry land. Each uses incredible jaw strength, speed, oversized paws and claws, and little fear of man to send their prey to its eternal reward . . . deep inside their stomachs.

Tigers and lions are stealth predators that typically chase and take down their victims by overpowering them. Once the prey—that would be you—is taken down, you are likely to be dispatched with a bite through the spine, windpipe, or jugular vein, or an immediate evisceration of your vital organs. Bears, especially polar and grizzly bears, kill in a similar manner, although they favor prey found in water, such as seals and salmon. Their incredibly muscular shoulders and forearms allow them to knock around large prey, even a good-sized human such as yourself, with the ease of flipping pancakes.

Your final meeting with any of them is bound to occur when you intrude on their turf, especially if you're threatening their cubs or are in the way of a food source. It will then be followed by a very brief run in which you are taken down and dismantled by teeth and claws.

KNOWN BY SCIENCE AS:

African lion (*Panthera leo*), tiger (*Panthera tigris*), polar bear (*Ursus maritimus*, from the Latin "maritime bear"), grizzly bear (*Ursus arctos horribilis*, Latin and Greek meaning "bear-bear horrible"). Lions and tigers are distant relatives; bears are a genus unto themselves.

MEDICAL CAUSE OF DEATH

Blood loss; organ failure; trauma

TIME TO KILL

Typically less than ten minutes, as these predators often mortally wound their prey and then wait patiently until all signs of life have spilled out in a massive puddle of blood.

HIGHEST RISK

Backpackers in North America; bushmen and ill-equipped tourists in Africa; villagers in Asia

LETHALITY

Exceedingly high. Going one-on-one with a lion, tiger, or bear is guaranteed to end in tears for the human without a weapon. Bears and tigers have especially strong paws that can decapitate humans, sever major arteries, or shear limbs with a single swipe. All three animals are able to shred flesh, organs, muscle, and tissue with teeth and claws. Even if an attack is interrupted, the proximity of dagger-like teeth and sharpened claws to human organs leaves little room for salvation after a mauling has begun. There is also the weight advantage (not to mention the hunger factor): 1,000 pounds for polar and grizzly bears, 600 pounds for tigers, 500 pounds for lions.

KILLS PER ANNUM

- Grizzly and Polar Bears: six to ten per year 
- Tigers: 50 to 100 per year
- African Lions: 50 to 100 per year

HISTORIC DEATH TOLL

Perhaps millions. There is no way to tally the number of African tribesmen, Asian villagers, and North American natives who have become hors d'ouevres for Earth's largest land predators. The first cave paintings created by humans show these predators in full-on attack mode, so it's a safe bet that they have been snacking on our relatives since way before anyone started keeping track.

NOTABLE VICTIMS

- Tiger: Carlos Sousa Jr. taunted a Siberian tiger named Tatiana at the San Francisco Zoo on December 25, 2007. The beast then vaulted the twelve-foot high walls of its enclosure to hunt Carlos down and maul him to death in a nearby café.

- Bear: Self-proclaimed "Grizzly Man" Timothy Treadwell spent his adult life attempting to show the world—especially schoolchildren—that grizzlies were of little threat to humans. He kept that shtick going right up until October 2003, when a grizzly bear dragged Tim and his girlfriend out of their tent and ate them.

- Lion: Two lions in the Tsavo River region of Kenya killed and ate more than 100 railroad workers during a nonstop massacre in 1898. Their story is recounted in the movies *Bwana Devil* and *The Ghost and the Darkness.*

HORROR FACTOR: 9 

We've come to expect these kinds of attacks throughout history, so we're rarely surprised by them. Ever since childhood, we've been told stories and fables of ravenous beasts that lick their lips in anticipation of a human breakfast. Besides, when a carnivore that weighs nearly half a ton is hungry—whether it's at the zoo, the circus, or out in the wild—an unprepared human has to figure that the results of any altercation are not going to end well for the two-legged combatant.

GRIM FACTS

- Tigers have hunted humans more aggressively than any other animal, eating more than 1,000 people annually in southeastern Asia during the nineteenth and early twentieth centuries.

- Humans kill more lions, tigers, and bears every year than the other way around.

- Estimates are that there are more than 10,000 lions and tigers kept as pets or housed in private residences in the United States.

- On August 13, 1967, two 19-year-old girls—camped ten miles apart—were killed during the night by two different bears in Glacier National Park.

MENINGITIS

The lining inside your skull is supposed to keep your brain nice and safe. But if it gets infected, your life will shatter like a Ming vase delivered by an angry U.S. mailman.

HOW IT KILLS

Your brain is wrapped in tissue called the meninges. It looks something like leathery wrapping paper and serves the same purpose as bubble wrap: it protects your brain and your spinal cord from the outside world. The meninges have three layers: (1) the dura mater, which helps keep the brain from moving around; (2) the arachnoid layer, which contains blood vessels that clean and filter liquid around the brain; and (3) the pia mater, which fits snugly around the whole brain like plastic shrink wrapping. The pia contains blood vessels that supply the brain with glucose and oxygen.

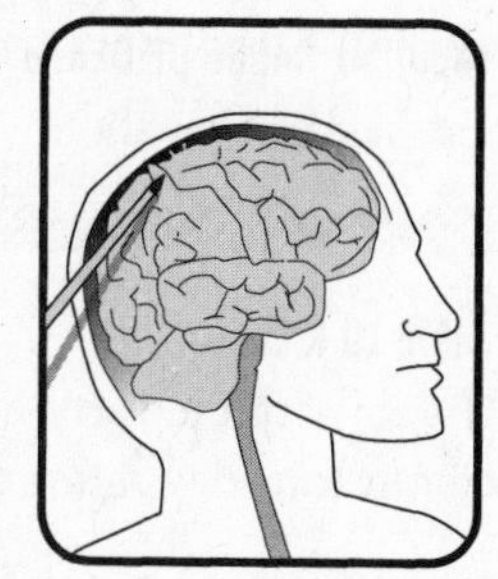

Infection and inflammation of the meninges is called meningitis. A number of things cause meningitis, ranging from skull injury to viruses, but the most lethal result of any initial cause is bacterial infection. The worst form of bacterial meningitis is meningococcus meningitis (or meningococcal disease), which is spread from person to person via saliva, sneezing, coughing, handshakes, kissing, and general close contact.

The first signs of meningococcus meningitis manifest themselves as a high fever and sensitivity to light, coupled with neck stiffness and headaches. Purple rashes or bruises will appear on your skin as your capillaries start to leak blood. As the bacteria works its way through your bloodstream and deeper into your brain, these symptoms will soon be followed by projectile vomiting and general disorientation and confusion. Seizures may follow and you'll have difficulty staying awake.

The infection will spread through what was once the protective covering of your brain, eating away at the blood brain barrier that keeps toxins out of your brain. Damaged blood cells will leak into your brain and clog your blood vessels, and cerebral edema (buildup of fluid in and around brain cells) will start drowning your brain. The resulting brain damage will cause seizures, and you will go into shock as your brain no longer has control over your blood flow. When your brain finally shuts down from this massive assault, so will you.

KNOWN BY SCIENCE AS:

Neisseria meningitides is the bacterium responsible for meningococcus meningitis.

MEDICAL CAUSE OF DEATH

Brain damage; shock

TIME TO KILL

Twenty-four to forty-eight hours after symptoms appear, which occur about four days after contracting the disease.

HIGHEST RISK

Anyone living in sub-Saharan Africa; babies in southeast Asia and South America; dormitory dwellers; high school kids sharing utensils and drinking fountains

LETHALITY

Medium to high. Without treatment, 70 percent of cases are fatal. If caught in time, meningitis can be treated. However, once symptoms show up, death may be the most desirable option as survivors frequently have severe brain damage, deafness, and retardation.

KILLS PER ANNUM

More than 150,000

HISTORIC DEATH TOLL

Millions. Large epidemics claim thousands of lives at a time; one 1996 outbreak in Africa resulted in 25,000 deaths.

NOTABLE VICTIM

Novelist and fabled man-about-town, Oscar Wilde, died of meningitis on November 30, 1900.

HORROR FACTOR: 2

It's hard to tell if you have meningitis, as the symptoms are flulike. Those who die from it are rarely diagnosed with it.

GRIM FACTS

- Meningococcal meningitis is one of many forms of meningitis (including viral, fungal, pneumococcal, and tubercular meningitis), but is especially dangerous due to the potential for its spread as an epidemic.

- The highest rate of meningococcal meningitis in the world occurs in a place called the "Meningitis Belt," which stretches across Africa from Senegal to Ethiopia.

- Meningitis has been caused in some cases by West Nile Virus, transmitted by mosquitoes instead of human-to-human contact.

MOSQUITO

The mosquito—an irritating, summer-spoiling, blood-sucking insect—is little more than an annoyance to you and me. This does not lessen the fact that mosquitoes are the most formidable killer in the entire animal kingdom, wiping out more than a million people every year.

HOW IT KILLS

The way a mosquito kills is quite simple: it sucks the disease out of one victim and injects it into another.

Female mosquitoes are hematophagous, which means they feed on blood (males do not). The females need the proteins in blood in order to nourish their eggs. To accomplish this, a female stabs her proboscis (a protrusion from the front of the head) into warm-blooded animals. Using her saliva as an anticoagulant, she then sucks blood out as needed. The bite is usually painless, and most people never realize they've been bitten until their bodies finally react to the saliva, thus triggering the well-known itch. By that time, the mosquito has already moved on.

The female mosquito flits to its next victim carrying the blood of its previous host. If the blood contains parasites, the parasites will

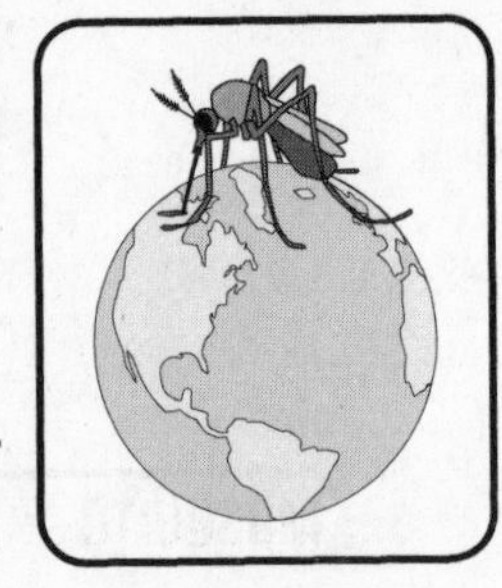

continue to live inside the mosquito and then be injected into a new host via the mosquito's saliva. In concept, it's very similar to contracting diseases by sharing dirty or used hypodermic needles.

And what a menagerie of diseases the mosquito can share! It is the primary carrier of malaria, a disease which breaks down red blood cells and can block the flow of blood to the brain. It carries the Flavivirus, a nasty virus that in various iterations is responsible for Dengue Fever (causes internal bleeding), West Nile Virus (causes brain disease), Yellow Fever (causes the vomiting of blood and coma), and Eastern equine encephalomyelitis (causes brain and spinal cord disease).

The only way to avoid getting killed by a mosquito is to not get bitten by one carrying a lethal disease.

KNOWN BY SCIENCE AS:

While there are more than 2,000 species of mosquitoes, the one with the highest kill count is *Anopheles gambiae*, which transmits the *Plasmodium falciparum* parasite responsible for malaria. The term mosquito is Spanish for "little fly."

MEDICAL CAUSE OF DEATH

Brain disease; kidney failure

TIME TO KILL

Symptoms related to mosquito-borne diseases can start after a week, and last weeks and even months until death occurs.

HIGHEST RISK

Children in Africa and Asia; tropical explorers; anyone south of the equator who sleeps without mosquito netting

LETHALITY

Very high. You can kill a few mosquitoes at a time, but there are uncountable numbers right behind them . . . and they'll keep attacking.

KILLS PER ANNUM

More than a million people die each year from mosquito-borne diseases. Some estimates are as high as five million.

HISTORIC DEATH TOLL

Billions. Do the math: a minimum of a million people a year going all the way back to ancient times.

NOTABLE VICTIMS

British poet George Gordon Byron, better known as Lord Byron, died from malarial fever on April 19, 1824. Oliver Cromwell, soldier, statesman, famed "Roundhead," and leader of England, died from malaria on September 3, 1658. Italian poet Dante Alighieri, the creator of the *Divine Comedy* and the fabled "Inferno," died of malaria on September 14, 1321.

HORROR FACTOR: 1

Mosquito bites aren't all that bad until the actual disease kicks in.

GRIM FACTS

- Mosquitoes are considered a primary "disease vector," meaning they do not cause the actual disease, but transmit and spread the disease by carrying it from one host to another.

- Attempts at mosquito eradication on the part of governments and health organizations have failed miserably. Incidence of mosquito-borne diseases can be radically reduced by using mosquito netting and screens.

- Adult mosquitoes live from one to three months.

- Mosquitoes don't transmit AIDS because the human immunodeficiency virus (HIV) is actually digested and destroyed inside the mosquito.

Game Over

Videogame addict Lee Seung Seop suffered heart failure brought on by dehydration and exhaustion on August 5, 2005. The twenty-eight-year-old Lee had just finished playing fifty straight hours of the battle game *StarCraft* at an Internet café in South Korea.

NATURAL DISASTERS

Hey God, what's with the attitude?

HOW IT KILLS

Tornado. Landslide. Earthquake. Monsoon. Volcano. Hurricane. Tsunami. Typhoon. Avalanche. Clearly our home planet is trying to get rid of us. We classify such events as "natural disasters" even though the damage is measured in loss of human life and destruction of man-made structures. The disaster isn't happening to nature—in fact, for nature itself, earthquakes and volcanoes are really no biggie. The planet will spin on long after we're gone whether it's covered in lava, or daisies, or Styrofoam. We care what happens to the Earth when it happens to us.

And so it's not without a touch of hubris that we build homes on cliff sides, cities on fault lines, and farms in flood zones. Given our growing numbers, coupled with the unpredictability of nature, having humans live in vulnerable areas can hardly be avoided. There's simply not a place on the planet insusceptible to nature's ever-changing moods. Still, we can make some population-saving choices of our own, like not building a village of huts under a mountain that's still smoking. We can develop weather-alert systems,

plan escape routes, board up windows, and reinforce levees. But in the end, when Mother Nature really wants to come get you, you're gonna get got.

Natural disasters are horrific on a biblical scale—they can pluck human lives by the tens or even hundreds of thousands. In the worst catastrophes, the dead are so impossibly numerous and the local infrastructure so debilitated that assessing the human toll is left to guesswork. Estimates on the 2004 tsunami in the Indian Ocean, for example, still range from 275,000 to more than 350,000. After the Huang He River in China caused extensive floods in 1931, the dead numbered somewhere between 800,000 and 4,000,000 (a range of 500 percent); the only statistic that researchers generally agree on is that it was the worst disaster of the twentieth century.

Asia in particular is accustomed to its role as God's punching bag. Dense populations, geographic vulnerability, and lack of resources for preparedness and recovery make for a perfect storm of disaster conditions in several Asian regions. A 1970 cyclone that hit the Ganges Delta in present-day Bangladesh killed half a million people; the 1976 earthquake in China's Tangshan region left 655,000 dead; in 2004, 275,000 were killed in the Indian Ocean wildfires.

If you were snuggled up for sleep inside a mountainside ski lodge (say, a lodge in Thredbo, Australia), your biggest concern might be where to find hot chocolate in the morning. But suddenly you hear a roar that sounds like fighter jets are landing on the roof. You don't know it, but one hundred tons of earth has shifted off the mountain some 325 feet above your bedroom. And riding on top of the landslide is another lodge, already collapsed. In seconds, the earth and the shattered structure on top of it cascade down the 70-degree slope and slam into the wooden lodge where you lay inside. Before you can even throw back the covers, the wall behind your head folds like it was made of paper—and in crashes an enormous heap of dirt, earth, and wooden beams. In the split second you have to take it all in, you glimpse the broken bodies of lodgers who'd been sleeping in the uphill building, and then you're engulfed in the horrid mass. Tumbling now as if you were thrown inside a clothes dryer, your body is being crushed and broken by the fast-moving mountain of earth and flotsam. The last things you see before all the air is squeezed from your lungs are a mud-covered television, a no-parking sign, and a final glimpse of the clear night sky.

Every region has its specialty, whether it's tornadoes in the Midwestern states, landslides in the Philippines, or volcanoes in central America. Stay tuned to the disaster channel. There's always more to come.

KNOWN BY SCIENCE AS:

Absolute disaster

MEDICAL CAUSE OF DEATH

Various, though dominated by traumatic crush injuries and drowning

TIME TO KILL

Seconds, then days or weeks for stranded victims who can't find clean water or fulfill basic health requirements

HIGHEST RISK

Poor residents of heavily populated cities that are prone to flood

LETHALITY

Extremely high, but dropping. The number of people affected by natural disasters has risen sharply over the past forty years while the number of global fatalities has been steadily dropping for an entire century.

KILLS PER ANNUM

Varies greatly, since a single disaster can triple the number of fatalities from a previous year. According to a report by the United Nations, 16,517 people were killed by natural disasters in 2007. In the previous year, the mortality statistic was higher by 5,000.

HISTORIC DEATH TOLL

Untold billions. In league with disease and pestilence, disasters represent the most effective means of keeping the world's population in check.

NOTABLE VICTIMS

In the aftermath of the 2005 Hurricane Katrina, thirty-nine-year-old Joseph Major did not want to abandon his family home. He was found wearing a rubber bracelet reading "Save America's Wetlands." Major is listed among Katrina's 1,294 deceased. Another 595 are still listed as missing.

HORROR FACTOR: 8

The scope and intensity of devastation makes for a living Hell on Earth. And if you make it to heaven, you just know there's going to be a line at the gate.

GRIM FACTS

- The indirect effects of a disaster can wipe out scores of victims long after the immediate threat has passed. People with chronic conditions are unable to get their medication; structures are prone to collapse; water sources become contaminated with chemicals and swarm with disease.

- In the wake of Hurricane Katrina, survivors wading through New Orleans faced waters filled not only with sewage and chemicals, but also with rattlesnakes, copperheads, and water moccasins.

NERVE GAS (SARIN)

The term "nerve gas" elicits a primal kind of fear. These chemical weapons—best exemplified by sarin—throw the nervous system into overdrive in the pursuit of a complete and deadly kill.

HOW IT KILLS

Developed as a pesticide by German scientists, sarin is a man-made chemical that is designed specifically to kill. It is a clear, colorless, odorless, and tasteless liquid that is easily spread through the air in a gaseous form. It can also poison water and food, and be ingested. If you drink it, touch it, or breathe it, sarin immediately works its way into your nervous system. And we do mean immediately: it takes only seconds for sarin to start working. Like all nerve gases (also called nerve agents) chemicals block the "off" signals that the brain sends to muscles and glands, resulting in them becoming overactive and overstimulated. Your brain's traffic light is stuck—for as long as you might live—on bright green.

Immediately, your nose will start running and your eyes watering. Your vision will blur and your eyes will actually start to hurt as your pupils shrink down to the size of pinpoints. Your muscles will twitch uncontrollably, and sweat will begin pouring out of your body while drool flows out of your mouth. Your chest will tighten and your breathing will become rapid. The act of breathing will be made that much more difficult by the fact that you're also coughing.

Diarrhea follows as does the need to urinate. You'll get sick to your stomach and start vomiting. A headache will be followed by confusion and drowsiness. You'll start having convulsions, and your heart rate may speed up (or even slow down). As your muscles reach a point of massive overstimulation, they will stop functioning and you'll be paralyzed. You will then slip into a coma, cease breathing, and die.

KNOWN BY SCIENCE AS:

The word "sarin" was created by using letters from last names of its inventors (Schrader, Ambrose, Rudriger, and van der lINde). It is called GB by NATO (North Atlantic Treaty Organization): G for Germany, where it was invented, and B because it was the second nerve agent developed there.

MEDICAL CAUSE OF DEATH

Asphyxiation

TIME TO KILL

Symptoms start within seconds, and a high dose can kill within an hour

HIGHEST RISK

Fledgling terrorists trying to create sarin on their own; enemies of those same terrorists

LETHALITY ☠☠☠

High. There is an antidote, but it's not likely you're going to get to it in time.

KILLS PER ANNUM

Because sarin is used sparingly by terrorists, it does not have an annual average.

HISTORIC DEATH TOLL

Could be thousands, as sarin was suspected of being one of the gases used by Saddam Hussein on Kurdish villages. In addition, there are probably a lot of farmers and government volunteers who have been sent to their final rewards, thanks to sarin.

NOTABLE VICTIM

- In 1953, Royal Air Force engineer Ronald Maddison volunteered for a research program that he was told was looking for a cure for the common cold. An hour later he was dead. In 2004, a British court found that he had been given sarin as part of a government experiment.

- On March 20, 1995, members of the Aum Shinrikyo cult in Japan released sarin—from plastic bags—into the Tokyo subway on five different lines. Twelve people died, fifty more were critically injured, and more than 5,000 were admitted to hospitals. In a nod to the murder of Bulgarian dissident writer Georgi Markov by ricin, the cult members popped their sarin-filled bags with sharpened umbrellas.

HORROR FACTOR: 7

There is no way to immediately tell if you've been poisoned by sarin, so your initial response will be nil. This, however, will be offset by the realization that you no longer have control over your body functions.

GRIM FACTS

- Sarin was created by Gerhard Schrader and his coworkers in 1938. He was looking for a pesticide that would be so effective against insects that it might help eliminate world hunger. The Nazis appropriated it for use during World War II.

- The U.S. military stockpiled sarin canisters confiscated them from the Nazis. It is believed that some of them were disposed of in the ocean.

- Sarin and other nerve agents are often associated with mustard gas, in part because they are chemical weapons delivered in gaseous form. Mustard gas, though, is not a nerve agent—it causes skin blisters—and is rarely fatal.

- Nerve agents are considered "weapons of mass destruction" and were outlawed by the Protocol for the Prohibition of the Use in War of Asphyxiating, Poisonous or other Gases, and of Bacteriological Methods of Warfare, in 1928. This treaty is simply referred to as the Geneva Protocol.

NUCLEAR EXPLOSION

A new clear day will be your last day on Earth.

HOW IT KILLS

If you're anywhere near a nuclear explosion, make sure you're as near as you can get—preferably within five miles. That way you will die as quickly as is physically possible, while those farther away from the blast will suffer horrendous pain that may last for minutes, hours, days, and even years. Trust us, you want to get this one over.

Death from the actual explosion is significantly different from being killed by the fallout or radiation. Those have their own ways to kill you (*see* **Radiation**), mostly involving long-term cellular damage and mutation.

If you're close to the detonation point—the hypocenter or "Ground Zero"—things will happen very quickly and very powerfully. The blast from an exploding nuclear device creates a shock wave producing thousands of pounds of pressure per square inch (PSI). It also generates heat in excess of ten million degrees.

This then rushes toward you at a speed of more than 900 miles an hour.

The pressure from the shock wave is known as overpressure, because it is over and above the standard atmospheric pressure we live with everyday. The organs in your body can withstand overpressure of about 15 PSI, give or take a few pounds. When the blast hits you at anything over 30 PSI, your lungs and heart collapse immediately and your tissues tear apart. The attendant heat will melt your flesh, eyes, and then your internal organs. All your organs and cells will be incinerated and reduced to ash. This will happen before you can even blink your eyes, which were most likely blinded by the flash, anyway.

The only thing left of you will be your shadow burned into the concrete—which happened to the victims in Nagasaki and Hiroshima, Japan.

KNOWN BY SCIENCE AS:

Nuclear event. This event is caused by either joining atoms (fusion) or splitting them (fission) within an explosive device.

MEDICAL CAUSE OF DEATH

Organ failure; incineration; disintegration

TIME TO KILL

Immediate

HIGHEST RISK

People living near nuclear power plants; neighbors of do-it-yourself bomb makers; enemies of countries that have nuclear bombs

LETHALITY

The highest. If you're within three miles of Ground Zero, you will be dust in the wind within seconds.

KILLS PER ANNUM

Less than one, except in 1945 (see below)

HISTORIC DEATH TOLL

Approximately 120,000 people died almost instantaneously from the two bombs dropped in Japan in August 1945.

NOTABLE VICTIMS

Eighty thousand in Hiroshima on August 6, 1945, and 40,000 three days later in Nagasaki

HORROR FACTOR: 0 

Bright light. Death. The end.

GRIM FACTS

- When J. Robert Oppenheimer, the father of the atomic bomb, saw the first nuclear explosion test, he said, "Now I am become death, destroyer of worlds," a line adapted from the *Bhagavad Gita*, a sacred Hindu text.

- The atomic bombs dropped on Hiroshima and Nagasaki were approximately 15 and 20 kilotons, respectively. Current nuclear weapons are measured in the hundreds of kilotons, or megatons.

- After reaching a temperature of millions of degrees, the heat from a nuclear blast rapidly "cools" down to several thousand degrees.

- "Megadeath" is the name given to the aftermath of a nuclear explosion that results in deaths of a million or more people.

Paranoia + No Food = Death

Kurt Gödel, the renowned twentieth-century mathematician and philosopher, suffered from mental illness late in life and came to believe that someone was trying to poison him. To protect himself, Gödel relied on his wife to taste his food to ensure that it was safe. When his wife was hospitalized for her own medical condition—and unable to check his food—Gödel refused to eat anything at all. He died from malnutrition on January 14, 1978, and weighed just sixty-five pounds.

OLEANDER

A lovely and decorative plant used for shade in hot climates, the oleander's attractive appearance and pleasant scent are screens for a plant that is all too happy to break your heart.

HOW IT KILLS

Oleanders are found all over the world, primarily in subtropical and desert areas. They are evergreen bushes that grow in dense thickets, and can reach heights of nearly twenty feet. The oleander has long slender leaves like a bamboo, coupled with deceptively beautiful white, red, pink, and yellow-orange blossoms, making it ideal for natural fencing around homes and parks. The fact that oleanders need very little water also makes them a popular choice for desert landscapers who want to plant environmentally friendly plants.

That's about as friendly as the oleander gets. One of the few truly deadly plants on Earth, an oleander can kill you even if you don't eat it. Its leaves and branches are suffused with a sap that contains cardiac glycosides (glycosides are molecules that bind sugars to nonsugars), which interrupt the normal functioning of your heart. This creates arrhythmia that can force your heart to the extremes of a wildly increased heart rate or complete cardiac shutdown.

The sap can get into your body if you eat the leaves or if you're stabbed by a broken branch (murderers have made tea from the oleander). Several hours after being poisoned, stomach pain and vomiting begins. Your skin starts turning blue due to a deficit of oxygen in your bloodstream, your pupils dilate, and you get vertigo. Before you pass out, you'll get bloody diarrhea and have seizures. The last stage before the curtain falls will be respiratory paralysis, but ultimately it will all come down to your heart giving up on you.

KNOWN BY SCIENCE AS:

Nerium oleander, or common oleander, as well as *Thevetia peruviana*, called the yellow oleander. Other popular names include rose laurel, adelfa, and karavira. Oleandrin is the name for the principal glycoside in oleanders.

MEDICAL CAUSE OF DEATH

Heart failure; respiratory paralysis

TIME TO KILL

One to two days

HIGHEST RISK

Kids playing inside the dense branches; unprotected and uninformed gardeners; burglars trying to escape through its fencelike fortress

LETHALITY

Low. Oleander poisoning is survivable, as long as you know you've been poisoned. The treatment for those who have eaten or swallowed parts of an oleander relies on the tried-and-true methods of stomach pumping or vomiting. Oleandrin can also be eliminated from the body by having the victim ingest activated charcoal, which binds with the glycoside and forces it from the body.

KILLS PER ANNUM

Several hundred to a thousand, mostly in Sri Lanka

HISTORIC DEATH TOLL

Some reports have claimed a historical death toll of 2,000 to 3,000 deaths per year for the past two millennia, resulting from the big three of accident, homicide, and suicide. That's roughly 3 to 6 million people.

NOTABLE VICTIMS

The mysterious deaths in May 4, 2000, of adopted Russian boys Alexei Wiltsey, two, and his three-year-old brother, Peter, in Los Angeles were ultimately attributed to the boys having eaten oleander leaves. It took the L.A. County coroner's office two months to determine the cause of death after authorities originally attributed it to rare diseases contracted in the boys' homeland.

HORROR FACTOR: 1

It's not really all that bad, when you think about it. You get a splinter, or you eat a nice-looking blossom. Two days later, you're in a pine box. It doesn't get much less melodramatic than that.

GRIM FACTS

- Deliberate ingestion of yellow oleander seeds has become a popular twenty-first-century method of committing suicide in Sri Lanka. The seeds are known as "lucky nuts," and thousands of cases are reported every year, with a fatality rate of nearly 10 percent.

- An urban legend perpetuates the story of an entire Boy Scout troop dying after toasting marshmallows on oleander branches during a campout.

- Extracts of oleander have been the focus of research to treat cancer, and mild doses have been used as folk remedies to ease menstrual pain. However, we don't recommend you try this at home, ladies.

PEPTIC ULCER

When you say your stomach is killing you, you might really mean it.

HOW IT KILLS

Peptic ulcers are sores in the upper digestive tract—small areas where the soft lining of the stomach, small intestine, or esophagus is open and inflamed. Most often they are characterized by a painful sensation in the gut, like something is trying to gnaw its way out of your abdomen.

Well, something is. The bacteria *Helicobacter pylori*, or H. *pylori*, is known to cause most peptic ulcers, and it does its work by invading the soft layer of protective mucous that lines your stomach and small intestine. Powerful stomach acid then erodes the exposed tissue, broadening and deepening the sore. That's when things start to hurt deep inside you. An untreated peptic ulcer can perforate your stomach, meaning a hole is eaten right through your stomach lining. Your abdominal cavity, now open to the contents of your digestive tract, becomes susceptible to life-threatening infection. The raw

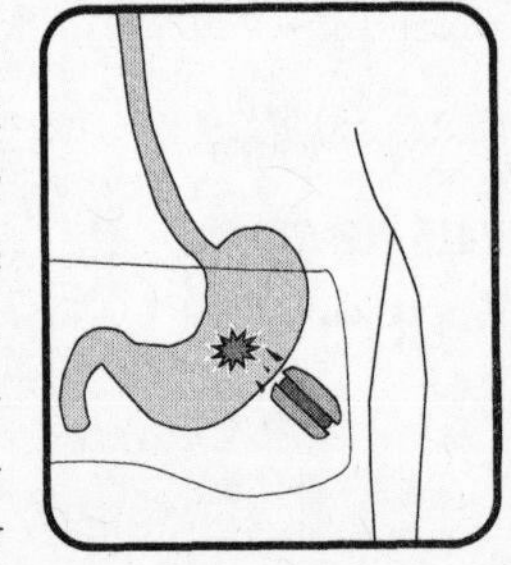

sewage, as it were, that spills into your insides can kill you as surely as if you were drowning in a septic tank.

Treatments are available for killing the *H. pylori* bacteria and reducing acid. Surgery can usually cure a problem that medicines cannot. Thanks in large part to the recognition of *H. pylori* as a culprit, fatalities are on the decline . . . though thousands in the United States still die every year of complications from perforated ulcers.

KNOWN BY SCIENCE AS:

Peptic Ulcer Disease (PUD); duodenal ulcer; gastric ulcer; esophageal ulcer

MEDICAL CAUSE OF DEATH

Infection

HIGHEST RISK

Elderly people with an *H. pylori* infection. Smoking, abusing alcohol, and overuse of nonsteroidal anti-inflammatory drugs like aspirin and ibuprofen can also cause PUD.

LETHALITY

Low, unless untreated. About 1 percent of the 400,000 people hospitalized for PUD die of complications.

KILLS PER ANNUM

Approximately 4,000

HISTORIC DEATH TOLL

About 425,000 in the past century. A 2001 study concerning England and Wales showed that the generation born around 1885 carried the highest risk of gastric and duodenal ulcers, with the heaviest toll on life occurring in the 1950s.

NOTABLE VICTIM

Filipino film idol Rudy Concepción died of a peptic ulcer at the age of twenty-eight while he was filming *Mahal Pa Rin Kita* in 1940.

HORROR FACTOR: 3

Knowing that abdominal pain will not be relieved by a trip to the bathroom, but instead may kill you while you're in there, earns PUD modest horror honors.

GRIM FACTS

- Stress does not cause ulcers, nor does eating spicy foods.
- *H. pylori* infection has been linked to instances of gastric cancer, lymphoma, and pancreatic cancer.

A Nightcap

Tennessee Williams, the Pulitzer prize–winning playwright who wrote *A Streetcar Named Desire* and *Cat on a Hot Tin Roof,* was in the habit of holding the cap to his eyedrops clenched between his teeth as he administered the drops into his eyes. On February 23, 1983, he was doing exactly that when he tipped his head back, and the cap slipped into his mouth, and he choked to death on it. Some have speculated that Williams's liberal intake of alcohol and pills may have prevented him from coughing up the cap.

THE PLAGUE

The plague is used as a catch-all term to describe epidemics. But there is a real disease called the plague, which is so hideous that you want to avoid it like the . . . well, you get the idea.

HOW IT KILLS

The bacteria that causes the plague lives in rodents and is carried between them by fleas. When the fleas jump to humans, the horror begins. When the fleas bite their human hosts, one of two kinds of plague bacillus is injected into the flesh. The first is the bubonic plague, which attacks the lymph glands. The bacteria travels through the lymphatic system until it reaches the lymph nodes. The nodes become inflamed, and swelling occurs in the armpits, neck, and groin. This appears on the skin as buboes: blotchy, red, large, and painful manifestations of the diseased lymph nodes. If this is what you contracted, your odds of dying are about 60 percent.

The second kind of plague that might be sent into your innards by a flea bite is septicemic plague. This goes straight to your bloodstream and attacks the entire body by riding along the path of the circulatory system. It attacks tissue throughout the body and causes internal bleeding, resulting in the coughing up of blood. This is

accompanied by both a fever and the chills, as well as stomach pain and shock. If you get this one, you're dead in twenty-four hours.

There is a third kind of plague, and it requires no fleas to pass it around. This is pneumonic plague, and the bacillus lives in your lungs where it destroys tissue. Your lungs fill up with liquid, resulting in extremely severe pneumonia. The bacteria live in the liquid in your lungs and are then passed from person to person during coughing fits that spew water droplets into the air and into your loved ones' own respiratory systems. You'll only have to watch them cough up their guts for three days, though, because that's all the time you have left to live.

In all cases, the plague causes internal bleeding right under your skin, causing your flesh to turn a purplish black just before you die. This grotesque discoloration gave the famed "Black Death" (a pneumonic plague) its name.

KNOWN BY SCIENCE AS:

Plague is caused by the *Yersinia pestis* bacteria.

MEDICAL CAUSE OF DEATH

Respiratory failure; blood loss

TIME TO KILL

- Bubonic: one to eight days
- Septicemic: one day
- Pneumonic: two to four days

HIGHEST RISK

Peasants living during the Dark Ages; pet owners in the U.S. Southwest handling infected pets

LETHALITY

High. Each form has a different lethality, averaging between 50 and 89 percent, although each can be treated if caught in time. Bubonic is the most survivable, septicemic slightly less so. Left untreated, pneumonic is invariably a one-way ticket to the morgue. Or the funeral pyre.

KILLS PER ANNUM

More than 150, mostly in Africa

HISTORIC DEATH TOLL

Countless millions. During the 1300s, the "Black Death," a colorful form of the plague, killed 25 million people in Europe, one-third of the extant population. It did even more damage in Asia, where it was estimated to take out twice that many people. And the Great Plague of London in 1665, took out nearly 100,000 people in just six months.

NOTABLE VICTIM

Erik York, a thirty-seven-year-old wildlife biologist working in Arizona's Grand Canyon National Park, died of the plague on November 2, 2007. He contracted the disease after performing an autopsy on a mountain lion. York and his coworkers thought he merely had the flu.

HORROR FACTOR: 9

The horrid prospect of dying from the plague is multiplied by the fact that if you're dying from it, probably most of the people you know are, too.

GRIM FACTS

- The western United States has annual reports of plague, mostly found in animals such as desert rodents like prairie dogs and rock squirrels. New Mexico, however, has reported more than 250 cases of human plague since 1949, with thirty-three deaths.

- An average of thirteen human plague cases are reported annually in the United States, while the worldwide total is close to 3,000 every year. The vast majority of these are nonfatal.

- The largest recent plague epidemic was in India in 1994. Fifty-two people died and more than a quarter million people left their homes in fear of catching the disease.

PLAYING PROFESSIONAL SPORTS

None of the mainstream American sports like baseball, basketball, and football qualifies as the Most Dangerous Game. Even in hockey, where players are routinely bludgeoned by fists and sticks, deaths are rare. But plenty of other athletes risk life and limb at professional play.

HOW IT KILLS

Getting killed in a boxing match isn't officially an objective of the sport, though anyone watching a vicious bout might say we're splitting hairs. It's one of several games that truly can end in sudden death. And, no, there won't be any overtime.

Pugilist pros are accustomed to getting all kinds of things knocked out of them, including their teeth, their sense, and occasionally their lives. When a boxing opponent delivers a hook to your head, the blow approaches a force of .63 tons—which the American Medical Association has likened to a thirteen-pound wooden mallet being swung at 20 miles per hour. Your neck twists and your body follows as it reels from the punch, and the delicate blood

vessels and nerve fibers that attach your brain to the inside of your skull are torn.

See, the problem is that the soft tissue of the brain can move a bit within the cranium; with a heavy blow, your whole cranium spins while your brain momentarily stands still. That's not good. Before you even begin to feel the pain of the blow, your cognitive abilities are stupefied, you can't speak or stand, and you begin to black out as your brain hemorrhages. A single punch can be fatal, and repeated batterings have an effect similar to crashing your car into a tree—provided the tree reached out and beat you with a branch several times after impact.

Whereas boxing fans are shocked when a fighter loses his life in the ring, the draw of professional auto racing is in seeing the drivers cheat death lap after lap. The dirty little secret of this sport is that spectators are thrilled at the prospect of witnessing a spectacular crash: the number of TV and live viewers tracks directly to accident statistics. Speeding toward the checkered flag like it was the gates of St. Peter, more than 180 drivers have been killed on the world's five most dangerous tracks alone. Many die like Dale Earnhardt Sr. did in 2001, when his Number 3 car hit the wall at a speed estimated to be 170 mph. "The Intimidator" was pronounced dead with a fatal skull fracture.

At least race-car drivers have tons of steel to protect them in a crash. Not so for those riding high in the saddle. Horse racing and even horse jumping are among the world's most dangerous sports. Known in equestrian circles as show-jumping, the sport can seem quite elegant and elite—until the horse stumbles. Suddenly, he's riding you instead of the other way around. Most "eventing" fatalities are the result of rotational falls, where hitting an obstacle sends the horse and its rider into a fatal somersault. Halfway through the rotation, you and your pal Seabiscuit are airborne and facing skyward. Then a ton of horse—literally, a ton—lands on your chest. To put it in technical terms, you get squished. The human frame is designed to sustain your own weight, not yours plus Mr. Ed.

KNOWN BY SCIENCE AS:

Game over

MEDICAL CAUSE OF DEATH

Subdural hematoma (boxing); basilar skull fracture (Dale Earnhardt's racing crash); multiple crush injuries (show-jumping)

TIME TO KILL

Usually less than one minute

HIGHEST RISK

Auto racers at Indianapolis Motor Speedway, the world's deadliest racing track; anyone in the general vicinity of Mike Tyson

LETHALITY

Low, according to professional sports organizations. With thousands of pros competing in dozens of dangerous games every year, the risk of dying in any single event is below .1 percent.

KILLS PER ANNUM

Boxing and show-jumping fatalities *each* average more than ten per year. NASCAR racing averages four per year.

HISTORIC DEATH TOLL

More than 900 professional boxers have died in the past ninety years as a result of injuries sustained in the ring. That figure doesn't even include amateur bouts, where headgear is required but doesn't necessarily keep your brain from sloshing around inside your skull.

NOTABLE VICTIMS

Back in 1924, boxer Ralph Thomas had a modest professional record of 0 wins and 1 loss. When a fighter didn't show up for a bout with middleweight Alberto Icochea, Thomas volunteered to fill in. During the second round, Thomas took a hard hit under his heart and collapsed on the mat, never to revive. He was now 0 and 2, lifetime.

Professional baseball players generally don't need nine lives to survive nine innings, with just one exception. In 1920, shortstop Ray Chapman of the Cleveland Indians died after taking a pitch in the head. The ball smashed a small section of skull, causing blood clots in his brain, and he died twelve hours later. Yankees pitcher Carl Mays heard Chapman's skull crack, but, thinking it was the ball against the bat, completed the play to first base.

HORROR FACTOR: 2

Nobody gets in the game to die but they ultimately go to the field of dreams doing what they love.

GRIM FACTS

- Brain hemorrhage is just one of the ways to be TKO'd in a boxing bout. Pros have died of kidney failure, heart dilation, neck fracture, and other causes under the bright white light.

- Soccer players periodically drop from preexisting heart conditions or from running into goal posts, but the sport loses far more fans worldwide in riots and bleacher collapses. When fans balked at a referee's call in a 1964 game between Peru and Argentina, the resulting riot killed 318 people.

- Horses fare worse than their jockeys and riders do. From 2004 to 2008, 3,035 horses died of various causes at racing facilities.

PNEUMONIA

Americans talk about pneumonia as if it was little more than a very bad flu. We might take it more seriously if we realized that it kills more than three million people every year, including nearly 100,000 people in the United States alone.

HOW IT KILLS

Pneumonia is an inflammation of the lungs, which can be caused by a number of things including chemicals, parasites, viruses, and fungi. Bacterial pneumonia is the most lethal, and you are most likely to get it by breathing in water droplets infected by someone who already has pneumonia. Most people who get pneumonia have weakened immune systems from something else such as a cold or chronic illness.

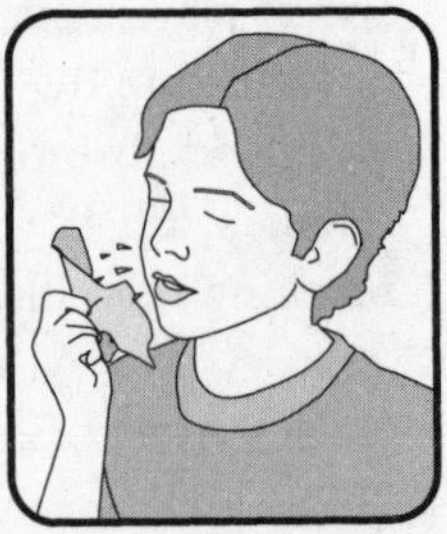

Once you've inhaled a particularly virulent strain of bacteria, such as staph or strep, it travels down to your lungs where it assaults your alveoli, the tiny air sacs in your lungs. The alveoli transfer oxygen into your bloodstream, but if they become inflamed by bacteria or a virus, they fill up with fluid and mucous, preventing the intake of oxygen. The more fluid that builds up, the more difficult it is to breathe.

In order to satisfy your body's need for more air, you'll start breathing heavily. The infection will result in chills and a high fever as your body tries to stop the spread of the disease. You'll start coughing regularly, and accompanying that cough will be lots of colorful mucous working its way up from your lungs. This might be green, yellow, or brown—and even red if there's blood in it. This rainbow assortment of sputum will be the only cheery aspect of having pneumonia.

Chest pains, from both coughing and the internal infection, will be a regular part of your daily routine. Vomiting and nausea will join the roster of ailments, as will diarrhea. Profuse sweating will highlight the fact that your lips and other areas of your skin are turning blue from lack of oxygen in your blood (adding another hue to the color palette of the disease). Eventually, you'll feel disoriented and regularly be confused about what's going on around you . . . which will help when you realize you can no longer breathe.

KNOWN BY SCIENCE AS:

Pneumonia is the medical term for the disease. *Streptococcus pneumoniae, Staphylococcus aureus, Haemophilus influenzae, Chlamydia trachomatis, Mycoplasma pneumoniae,* and *Legionella pneumophila* (Legionnaire's disease) are all bacteria that can cause pneumonia.

MEDICAL CAUSE OF DEATH

Respiratory failure. Pneumonia itself is actually listed as a cause of death on many death certificates.

TIME TO KILL

After infection, symptoms show up anytime from one to seven days later. Death can occur anytime from two to eight weeks.

HIGHEST RISK

Elderly people living in close quarters with others; patients in hospital surgical wards; children in Third World countries.

LETHALITY

Medium, about 20 to 50 percent. Vaccines and antibiotics have reduced the potential lethality.

KILLS PER ANNUM

More than three million, with some estimates as high as five million—the vast majority of whom are children

HISTORIC DEATH TOLL

If you do your math right, you come up with several billion just since the days of Julius Caesar.

NOTABLE VICTIMS

Pneumonia has taken more than its share of famous victims, including Franz Liszt, William Henry Harrison, Fred Astaire, Charles Bronson, and James Brown.

Muppet creator and *Sesame Street* icon Jim Henson became a famous victim of pneumonia when he died suddenly on May 16, 1990. Henson thought he had a cold he couldn't shake and after it worsened he finally checked himself into a New York hospital. He died within twenty-four hours after being admitted. By the way, septicemia—which builds on top of other diseases—was a contributing factor.

HORROR FACTOR: 3

It starts out as a cold and just gets worse. No really grim symptoms, such as seizures or internal bleeding.

GRIM FACTS

- Prior to 1936, pneumonia was the leading cause of death in the United States, and has remained in the top ten ever since.
- The flu is caused by a variety of influenza viruses, and in severe cases can lead to pneumonia. The two diseases are often linked in research and called P&I (pneumonia and influenza).
- Pneumonia kills more children worldwide than any other disease.

Gunsicle

On January 22, 1987, Pennsylvania state treasurer R. Budd Dwyer called a press conference to discuss his conviction for accepting kickbacks on state contracts. Facing up to fifty-five years in prison, and with the television cameras rolling, Dwyer pulled a .357 revolver out of a manila envelope, placed it in his mouth, and pulled the trigger. He died, as they say, "live and on the air."

POISON DART FROG

Coming in bright blues, greens, reds, oranges, and yellows, with splotches, stripes, and even polka dots in a little body about the size of a AA battery, the poison dart frog simply oozes cuteness. It also oozes the deadliest animal poison in the world.

HOW IT KILLS

Like most frogs, those of the poison dart variety are not interested in killing humans. They simply try to protect themselves from all predators by secreting poison through their skins. This poison is known as batrachotoxin, which is a strong neurotoxin that behaves like a cardiotoxin. While neurotoxins affect the functions of the nervous system, cardiotoxins go straight for the heart, and batrachotoxin does just that. The heart is set on a course of arrhythmia and fibrillation to the point that the only result will be heart failure. And it doesn't take much to kill you—about the amount of two grains of sand.

You might decide to pick up one of these little poison darlings while visiting some Amazonian outpost in Colombia. Sensing that you might want to hurt it, the frog releases its poison from glands and slicks up its skin. You touch the liquid, and as long as you have no cuts or nicks in your flesh, you're okay. But in this instance, you happen to have a small paper cut that leads right from your fingerprints to your bloodstream. The batrachotoxin is now heading straight for your heart, and the little frog really couldn't care less.

It causes heart failure rather quickly by throwing the heart's operation into immediate chaos, which spares you the kind of creeping paralysis that comes with other neurotoxins—so there's that to be thankful for. And the frogs provide a picturesque final image for you to view during your final moments.

KNOWN BY SCIENCE AS:

Phyllobates terribilis, the golden poison dart frog, is the most toxic creature in the animal kingdom. Some poison dart frogs are classified in the genus *Dendrobates*, but these are not noted as being of the ultra-deadly variety.

MEDICAL CAUSE OF DEATH

Heart failure

TIME TO KILL

Pretty quick, less than half an hour

HIGHEST RISK

Amazonian tribesmen; pet owners who haven't done all the background research on their recently cleansed poison pets

LETHALITY

High. If you've touched a poison dart frog, and its poison is in your bloodstream, you're already dead.

KILLS PER ANNUM

Not known. Since most exposure to the toxic frogs occurs among Colombian tribesmen who don't write, let alone keep medical records, statistics are nonexistent.

HISTORIC DEATH TOLL

Again, we're working with near prehistoric record keeping here, so any guess is probably a good one. We're estimating many thousands.

NOTABLE VICTIM

The first guy to pick one up

HORROR FACTOR: 2

You go from touching a cute little amphibian to absorbing the most lethal poison imaginable. That takes a little getting used to so it mitigates the overall horror.

GRIM FACTS

- One of the few things that reduces the effects of batrachotoxin is tetrodotoxin, the poison found in pufferfish.

- There are nearly 200 species of poison dart frog, which are also called dart poison frogs.

- Poison dart frogs get their toxicity from the insects they eat, converting chemical elements into incredibly lethal toxins. When these insects are removed from their diet, the frogs cease to be poisonous and are often kept as pets.

POLONIUM POISONING

A little radiation can really heat up the Cold War.

HOW IT KILLS

Polonium 210 is an extremely hazardous and unstable radioactive isotope, with few practical uses outside of triggering nuclear weapons. It was Alexander Litvinenko, the exiled Russian agent, who put this metalloid on the map when his poisoning came to light late in 2006. While the rest of the world was awaiting cool new spy movies like *Casino Royale* and *The Bourne Ultimatum*, the image of a deteriorating Litvinenko glared from the news pages as if to say, "You've got to be kidding me."

Litvinenko was a former KGB agent widely believed to have been poisoned by Russian authorities after turning on the Kremlin. The radioactive element he ingested—allegedly, in a cup of tea—left nuclear traces in hotel rooms, offices, and airplanes from London to Hamburg to Moscow.

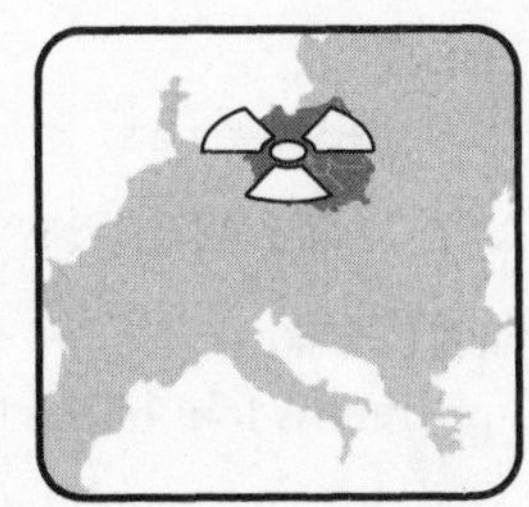

This was the first known case of deliberate polonium poisoning, but whoever asked Litvinenko to tea knew what fate a lethal dose of the radionuclide would bring.

Initially, it would have relatively minimal effects. But his condition would worsen faster than you can say лучевая болезнь (radiation sickness). Radiation accumulates in the body such that exposure after one month can be more than forty times what it was on the first day.

The first symptoms you'll experience after orally ingesting Polonium (Po) are nausea, malaise, and fatigue. Abdominal pain will cause you to retch and vomit. If exposure is high enough, you'll lose all the hair on your body, even your eyelashes.

Your response to Po poisoning will become more violent as the radiation goes to work on the primary cites of blood cell production (notably, bone marrow), resulting in blood infection and often fatal hemorrhaging. Your gastrointestinal system, however, is in for the worst blow. Severe diarrhea and intestinal bleeding will accompany a loss of fluids and electrolytes as segments of your GI tract blacken and die. Infected and bleeding internally, with cancerlike cell death gaining momentum in tissues throughout your body, you'll finally die.

KNOWN BY SCIENCE AS:

Polonium 210

MEDICAL CAUSE OF DEATH

Lethal exposure to radiation

TIME TO KILL

Twenty-three days

HIGHEST RISK

Apparently, foes of the Kremlin

LETHALITY 

Indisputable at high levels of exposure

KILLS PER ANNUM

Less than one

HISTORIC DEATH TOLL

One

NOTABLE VICTIM

Alexander V. Litvinenko died on November 23, 2006, and—as the number above attests—he is notable by dint of the fact that so far he's the only one.

HORROR FACTOR: 9 

Like an instantaneous, whole-body exposure to cancer

GRIM FACTS

- In your typical lethal dose of polonium, illness wouldn't set in for a week, with only mild initial symptoms following soon after. Litvinenko, who fell sick just one day after drinking laced tea, is believed to have received many times the lethal dose.

- Polonium was discovered by Marie and Pierre Curie. Having been named to publicize the independence of Poland, Marie's native land, the element's political implications come full circle with the Litvinenko poisoning.

Monkey Business

On October 20, 2007, while reading a newspaper on his apartment balcony, Surinder Singh Bajwa was attacked by wild monkeys. Trying to fend them off, Bajwa—the deputy mayor of New Delhi, India—fell over the edge and died the next day from his injuries. Up until his death, New Delhi officials had refused to address rising complaints from citizens about "the simian menace."

PUFFERFISH

All of the creatures in this book kill humans with deliberate intent, usually as a defensive measure but occasionally as a food source. The pufferfish is the only creature herein that kills humans only after humans have already killed it.

HOW IT KILLS

The fugu is commonly known as the pufferfish for its ability to blow itself up like a puffed-up ball. It is not, however, filled with air but with tetrodotoxin, one of the strongest neurotoxins on Earth. The poison keeps predators away, unless you happen to be a human eating the fish after it's already dead and sitting on the sushi table. At that point the tetrodotoxin in one pufferfish is enough to kill more than twenty full-grown people just like you.

In Japan, pufferfish is viewed as a delicacy, as long as the poison has all been—presumably—removed.

People pay a lot of money to eat it, believing it to be the most delicious of all fish, and sushi chefs who serve fugu have to be specially trained in how to remove the parts of the body—primarily ovaries and livers—that contain the neurotoxin.

So let's imagine your chef is having an off day and doesn't get the entire organ. Then let's imagine that you dip it in some soy sauce and wasabi, and swallow it down. At that moment, the game is underway. The neurotoxin attacks the nervous system in your body, preventing neurons from communicating with each other and with the brain. Most important, it stops the muscles from working.

It starts with tingling in your mouth, tongue, and lips. Dizziness, headache, nausea, and vomiting all are indications that things are not going to go your way from here on in. All motor functions along with autonomic functions (like breathing) shut down over the next several hours. You become paralyzed. However, you remain conscious while all this is going on, so you are fully aware that you are dying. Unless you are treated immediately with a respirator, you will die because there is no antidote. And your death gives new meaning to the phrase "dinner and a show."

KNOWN BY SCIENCE AS:

Takifugu rubripes, or tiger puffer, is considered the most poisonous of the two dozen species of fugu. Tetrodotoxin (or tetrodoxin) is named for *Tetraodontiformes*, the order of fish that includes fugu.

MEDICAL CAUSE OF DEATH

Respiratory paralysis; asphyxiation

TIME TO KILL

Twenty minutes to twenty-four hours

HIGHEST RISK

People who view eating sushi as an extreme sport on par with bungee jumping from skyscrapers

LETHALITY

High, upwards of 60 percent of all fugu poisonings end in death. Once it hits, you have less than an hour to get respiratory treatment to help you outlive the effects of the toxin.

KILLS PER ANNUM

Nearly 100 people, almost all of whom courted death by eating the world's most lethal delicacy

HISTORIC DEATH TOLL

Thousands, primarily in Japan and China

NOTABLE VICTIM

Mitsugoro Bando, considered one the greatest Kabuki actors of all time—and a man designated a "living national treasure" by the Japanese government—died of paralysis after eating fugu in a Kyoto restaurant in January 1975.

HORROR FACTOR: 9

Once it hits, you are SOL. Not a lot of time to do anything but realize you have only a few hours left on your life clock.

GRIM FACTS

- Tetrodotoxin is 1,250 times deadlier than cyanide. The amount equal to a pinhead will kill you.

- A famous Japanese saying regarding the transcendent taste of pufferfish states, "Those who eat fugu soup are stupid. But those who don't eat fugu soup are also stupid."

- Because tetrodotoxin victims are awake while they are dying, it is believed that pufferfish may be an ingredient used by Haitians to make "zombie powder."

RADIATION EXPOSURE

When the Geiger counter starts a-clickin', your clock is a-tickin'.

HOW IT KILLS

Like it or not, radiation is all around us, all the time. Radiation is a form of energy that travels in waves and, in some forms, has the power to penetrate materials like metal, fiberglass, or human tissue. There's radiation in the air we breathe, in the water we drink, and in the earth where we live. Even if you've chosen a career outside the field of plutonium production, you can't help but be exposed when you fly in an airplane or X-ray a tooth. Citizens of the Northeast, where radon naturally occurs in the ground, are annually exposed to the equivalent of two full-body CT scans. That pushes the level of safe exposure, but most of those residents grow old without ever sprouting a fin. If you're unfortunate enough to have moved in next door to a nuclear reactor, however, resist taking a dip in the neighbor's pool.

It's the threat of a significant radioactive event—an accident at a nuclear power plant, a dirty bomb, a nuclear detonation—that evokes Silkwood-scale paranoia in the average person . . . rightfully so, because radiation can penetrate the body and start messing with you on a cellular level. Depending on the type and extent of exposure,

radiation's raw energy can inhibit the reproduction of cells, cause immediate cell death, and disrupt the very composition of your DNA. That means certain types of radiation (namely, ionizing radiation and ultraviolet light) cause cells to mutate, and cell mutation is the hallmark of cancer. In the comic books, an acute exposure to radiation causes mutations that give you super powers. In real life, it's far less glorious than it was for Spider-Man.

Within an hour of severe exposure you'll suffer the first stages of acute radiation sickness, marked by nausea, vomiting, and diarrhea as your body tries to clear the poison. But it's in there too deeply—your cells have already been irradiated and the tissues of your organs are starting to die. Within a day you'll lose hair all over your body. Your temperature begins to rise even though your system has lost its ability to fight infection or heal wounds. Your digestive system is still expelling contents from both ends, but now the contents are bloody. Disoriented and dizzy, you lose consciousness as your blood pressure sinks from internal bleeding and the loss of blood through your bowels.

There is one upside: contrary to popular belief, you won't glow.

KNOWN BY SCIENCE AS:

Acute radiation syndrome; acute radiation sickness; radiation poisoning

MEDICAL CAUSE OF DEATH

Hypotension; blood loss (acute exposure); cancer (chronic exposure)

TIME TO KILL

Several days to two weeks with a single, huge dose of radiation absorbed. When exposure leads to the destruction of bone marrow, resulting in infections and internal bleeding, you may die after several months. At low but consistent exposures, cellular damage can lead to cancer and kill years later.

HIGHEST RISK

Enemies of any organization or nation with nuclear capabilities; residents living near test sites or faulty nuclear reactors; inhabitants of towns where nuclear fallout settles; emergency responders

LETHALITY

Very high. Radiation sickness is fatal in nearly all severe exposures.

KILLS PER ANNUM

Hundreds every year, considering the long-term effect of cancer in exposed victims

HISTORIC DEATH TOLL

In the hundreds of thousands, though accurate statistics are impossible to come by. Governments around the world are notoriously reluctant to release information about nuclear attacks, tests, and accidents.

NOTABLE VICTIMS

The World Health Organization (WHO) estimates that fifty-six people, most of them accident workers, died as a direct cause of the Chernobyl disaster in 1986. Another 9,000 living in nearby areas are believed to have contracted cancer as a result of the accident.

HORROR FACTOR: 6

It's a bloody, messy, painful death, with the only silver lining being that you won't survive to suffer through long-term cancer or to pass along your mutated genes.

GRIM FACTS

- The Mayak nuclear facility located in western Russia's Ural Mountains has been described as the most contaminated place on Earth.

- Even when radiation exposure is the origin of cancer, radiation therapy may be used to treat it. The same type of energy that causes cells to mutate can be used productively to prevent malignant cells from multiplying—ironically, by damaging them even further.

RED IMPORTED FIRE ANTS

In cowboy movies, the worst death a desperado could encounter was to be tied down under the hot desert sun, his body smeared with honey, and a busy nest of fire ants nearby. It's kind of comical in a Blazing Saddles kind of way until you realize just how painful a way it is to go.

HOW IT KILLS

Fire ants nest in the moistest soil they can find, so you're most likely to find them in yards or along sidewalks and under brick walks. Your personal encounter with fire ants will be as a result of disturbing the colony or threatening its queen, meaning you're standing on a nest, lying on one, or you're digging one up.

The nastiest of the fire ants is known as the red imported fire ant (RIFA), which was accidentally brought into the United States in the early 1930s aboard infested cargo ships from South America that docked in Alabama. Like all ground-dwelling insects that attack humans, the fire ant strikes wherever it can, which is going to be a toe, a foot, or a leg. Of course, it'll bite as high as it can get on your body before you swat it off. They are typically only about 5 millimeters long, which means you might not notice one until it bites you, but you're sure to notice when a lot of them bite.

The fire ant uses its mandible to grab your skin, but does not inject any venom from its mouth. Instead, once the mandible is locked into place, the ant punctures the skin with an abdomen stinger. It injects piperidine, an alkaloid poison, into the flesh. While the alkaloid is intended to kill the ant's prey—such as other small insects—in humans it produces a burning sensation at the point of entry. It results in a small whitish blister that rises from your skin. One sting is painful, but a host of ants all stinging at the same time is akin to lighting your flesh on fire.

If you don't have an allergic reaction, then you are simply going to experience a few days of painful swelling and scratching. If you have a reaction to the piperidine, however, you're in for a quick death by anaphylactic shock, meaning you won't be able to breathe and your blood pressure will plummet. And the ants will keep right on biting.

KNOWN BY SCIENCE AS:

The red imported fire ant is known as *Solenopsis invicta,* from the Greek "invincible." There are close to 300 species of fire ants worldwide, which have varying degrees of kill power.

MEDICAL CAUSE OF DEATH

Anaphylactic shock; heart failure

TIME TO KILL

From less than ten minutes to several months, depending on individual reactions. Some patients survive the initial reaction but fall into a coma.

HIGHEST RISK

People south of the Mason-Dixon line, especially kids crawling on a lawn and morons scuffing up ant hills while wearing sandals; people in wheelchairs; babies in parked strollers; and sleeping nursing-home patients

LETHALITY

Low. Less than 1 percent. A small percentage of those bitten and envenomated by fire ants experience anaphylaxis, and a fraction of those die. Antihistamines need to be ingested in order to reduce the possibility of anaphylactic shock.

KILLS PER ANNUM

Three to five, and increasing

HISTORIC DEATH TOLL

Since the 1930s, just under one hundred people have died from RIFA stings in the United States. Worldwide, the total is estimated to be in the hundreds annually.

NOTABLE VICTIMS

The Annals of Internal Medicine cited the deaths of two elderly nursing-home patients in Mississippi who were found covered with RIFAs in 1999. The patients had not disturbed the ants or their colony; instead the ants attacked them as they were asleep in their beds. One died after five days; the other held on for thirteen months.

HORROR FACTOR: 5

The odds of dying aren't too high, so you have plenty of time to put the situation in perspective and get a remedy. The creepy-crawly factor from having hundreds of these pests scurrying over your flesh is high, however.

GRIM FACTS

- The American Journal of Medical Forensic Pathology predicts that deaths from RIFAs will increase in the coming years as the species spreads across North America.

- The U.S. Food and Drug Administration (FDA) estimates that more than five billion dollars is spent every year on medical treatment, building damage, and control of RIFAs.

- In many parts of the world, the red fire ant is a non-native species that attacks animals that have no natural protection from their bite. It is now considered one of the most dangerous and destructive pests across the globe.

- Although she survived a RIFA attack in 1999, nursing-home resident Lucille Devers was awarded $5.3 million in damages by an Alabama court after being stung hundreds of times. Ordered to pay up were the assisted care facility and Terminix.

RICIN

The ubiquitous castor bean—grown all over the world and used as a flavor enhancer in foods—is the source for ricin, a favorite toxin of budding bioterrorists the world over.

HOW IT KILLS

The castor bean (which is really a seed) is used as a food additive and flavor enhancer by food manufacturers across the globe. However, the seed pulp or mash left over from the processing of the beans is rife with a toxin called ricin. Ricin causes cell breakdown in its victims, which leads to multiple organ failure throughout the body. What makes ricin especially nasty is that it is found all over the world and is readily available to just about anyone who wants it. The castor bean grows quickly in nearly all landscape settings, including home gardens, and grows at the rate of over a foot per month. Plus, it is estimated that some 50,000 tons of ricin-filled castor mash are processed each year. That's a lot of toxic leftovers.

Ricin has been weaponized as a powder, pellet, and mist by a number of nations and is designed to be inhaled, ingested, or injected. Depending on the form of contamination, you'll have a variety of symptoms, ranging from fever and weakness or fatigue to vomiting and bloody diarrhea. In all cases, your organs will shut down, leading to respiratory failure, renal failure, and pulmonary edema, all of which are followed by death.

The castor bean provides a singular example of the pain/pleasure principle that is part of all life and death. On one hand, it is the source of ricin, which is demonstrably one of the grisliest killers around. On the other, it is an essential ingredient in candy and chocolate, which are some of life's most indulgent pleasures. When it comes to the castor bean, you literally get to pick your poison.

KNOWN BY SCIENCE AS:

Ricinus communis is the scientific name for the castor bean plant, also called the palma Christi and the wonder plant. It can grow up to fifteen feet high in a single year.

MEDICAL CAUSE OF DEATH

Respiratory failure; pulmonary failure

TIME TO KILL

Symptoms can start within eight hours, and death will occur within two to four days.

HIGHEST RISK

Self-styled terrorists; children who mistake castor beans for cocoa beans; workers in castor oil plants

LETHALITY

High. There is no antidote known, which doesn't seem to bother anyone even though ricin is a by-product of food processing.

KILLS PER ANNUM

One or two, typically from accidental ingestion of castor beans or its extract

HISTORIC DEATH TOLL

Several thousand. However, weaponized ricin could take out hundreds or thousands of people during a single attack.

NOTABLE VICTIM

On September 28, 1978, Bulgarian writer and dissident Georgi Markov was killed in London after being jabbed by an umbrella. A pellet of ricin had been injected into Markov from the umbrella, and he died four days later. The pellet was found during Markov's autopsy.

HORROR FACTOR: 2

Because ricin's initial manifestations are similar to that of the flu, it won't be apparent to you—until way too late—how truly deadly it is.

GRIM FACTS

- Roger Von Bergendorff was hospitalized in Las Vegas, Nevada, on February 14, 2008, and went into a coma. Two weeks later, ricin was discovered in his hotel room. Initially, it was believed he might have been attacked—then authorities found out that Von Bergendorff may have created the ricin himself.

- One milligram of ricin is enough to kill an adult human being.

- Because of its ability to destroy cells—its cytotoxicity—ricin has been discussed by researchers as a potential "magic bullet" for targeting and killing cancer cells.

Dive Bomber

Soviet diver Sergei Chalibashvili attempted the incredibly difficult three-and-a-half reverse somersault in tuck position at the 1983 World University Games in Canada. The twenty-one-year-old Chalibashvili struck his head on the ten-meter-high platform, fracturing his skull. He went into a coma and died of heart failure a week later on July 25, 1983. For your information, the three-and-a-half reverse somersault in tuck position is officially known as the "Dive of Death."

SEPTICEMIA

You need your blood to be clean and to clean you out. When your blood starts carrying toxins around, it's time to call the mortician.

HOW IT KILLS

Septicemia is basically blood poisoning, and those are two words you really don't want to have next to each other in the same sentence. Septicemia often occurs after other diseases, such as meningitis, have already invaded your body's organs and tissues. Those infections can begin almost anywhere in the body, from the urinary tract to the meninges of the brain.

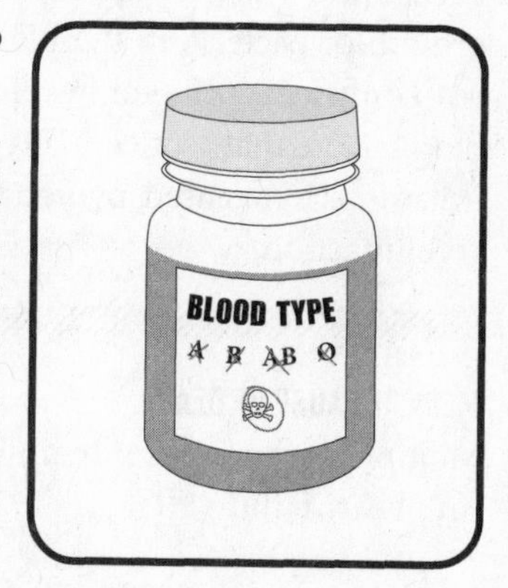

When the protective tissues or organs are eroded, these diseases spread to—and multiply in—the bloodstream. Blood vessels start spreading the disease, and the blood itself serves as the breeding ground for dangerous bacteria.

Once your bloodstream becomes infected, it's downhill from whatever other disease might have created it. You'll get a high fever, and lose interest in eating while your heart rate and breathing increase. You'll develop a rash that looks like bloody pinpricks on your skin (this is called a hemorrhagic rash). These little spots will spread until they look like bruises and then will merge together to form huge purple splotches on your flesh.

You'll get hypothermia, which is decreased body temperature, and your extremities will feel cold. You'll start feeling sleepy, but will have a general feeling of agitation, coupled with—and this is a medical condition—an overwhelming sense of impending doom. And you will also look unwell, another actual medical condition.

A strange clotting will occur in your blood that will start blocking your small arteries. The lack of blood flow to your organs will cause malfunctions, including the release of lactic acid into your blood and an inability to produce urine. You will likely go into septic shock, which is low blood pressure caused by the septicemia, and that will cause the subsequent failure of a multitude of organs, notably your kidneys and brain, but possibly your lungs and heart.

KNOWN BY SCIENCE AS:

Bacteremia with sepsis. Septicemia refers to the reproduction and growth of bacteria in the blood, while bacteremia—which it is often confused with—only refers to the presence of bacteria in the blood. Note that septicemia is a form of sepsis, which is whole-body inflammation caused by your immune system trying to fight a severe infection.

MEDICAL CAUSE OF DEATH

Multiple organ failure; brain damage; renal failure; respiratory failure; heart failure

TIME TO KILL

Very quickly. From the onset of septicemia—which is difficult to detect given that it usually arises from other infections—death can occur within forty-eight hours.

HIGHEST RISK

Hospital patients in intensive-care units; newborn babies; burn victims

LETHALITY

Medium, approximately 50 percent. Once septicemia becomes septic shock, it's all over.

KILLS PER ANNUM

Septicemia is not always separated from sepsis by health organizations, but a conservative estimate—based on U.S. deaths—puts the worldwide toll at more than half a million per year.

HISTORIC DEATH TOLL

Several million, most of whom died before the notion of clean hospitals became popular

NOTABLE VICTIM

Rupert Brooke, a popular World War I British poet, and at the time considered England's handsomest man, died of septicemia from an infected mosquito bite on April 23, 1915.

HORROR FACTOR: 8

Since you're probably already sick with something else, septicemia may not have much of an impact at first. But as soon as you stop pissing and your skin turns purple, you're certain to be extremely horrified.

GRIM FACTS

- Septicemia ranks in the top ten killers of people in the United States, accounting for at least 40,000 deaths per year.

- The bacteria responsible for the most common form of meningitis, *Neisseria meningitides*, often causes septicemia at the same time.

- Sepsis and septicemia come from the Latin *septicus*, which means putrefying or rotting. That, of course, is the same root for the term "septic tank."

SMALLPOX

One of the most widespread diseases in history, smallpox no longer scares the crap out of people like it did in days gone by. In fact, it has all but been eradicated from the face of the Earth. Fortunately, scientists have kept two samples of it alive to ensure that it is yet another hideous disease that will most assuredly kill you should it ever be liberated from the lab.

HOW IT KILLS

Smallpox is an infectious disease caused by the viruses *Variola major* and *Variola minor.*

You get smallpox by direct contact with someone who already has it, much like a cold. These people are contagious as long as they have the trademark "speckles"—a nice word for the rash pustules that scab over and are the signature image of the disease. You can also get it from bodily fluids associated with these people, as well as their clothing and bed linens.

Once you've breathed it in, smallpox takes about a week and a half to show symptoms. This starts with a fever in the 101 degree + range, accompanied by vomiting, fatigue, and overall body pain. A couple of days after this, red spots show up on your mouth and tongue, and work their way down your throat. Then they emerge as a rash on your face, which works its way over your entire body. This takes only about a day, and then the rash becomes a sea of pustules that rise up out of your flesh. Some of these pustules will form in and around your eye, causing corneal infection, which is likely to cause you to go blind.

The pustules may merge together to form what is called a confluent rash, turning large patches of flesh into a single toxic blister that eventually separates top layers of skin from lower layers. You may also begin bleeding under your skin. By now, you've had smallpox for more than two weeks, and you feel like you want to die. Well, you're in luck.

You are bound to die, especially if you're a small child, but the thing is that no one—to this day—really knows why. Scientists suspect that your immune system is simply overwhelmed trying to fight the massive viral attack on all parts of your body. Or it may be that the viral cocktail in your bloodstream eventually manages to shut down one organ after another once it has had its way with your now disfigured body.

KNOWN BY SCIENCE AS:

Variola major and *Variola minor, variola* referring to the spotting that occurs from the disease. *Variola major* has four primary strains (original, modified, flat, hemorrhagic), the latter two being inevitably fatal. *Variola minor* is a less common, and less lethal, form of the disease.

MEDICAL CAUSE OF DEATH

Immune system failure; organ failure

TIME TO KILL

Several days to several weeks, depending on your age and the form of smallpox. Most deaths occur by the sixteenth day.

HIGHEST RISK

Researchers and maintenance workers in the Centers for Disease Control (CDC) vault, and just about anyone downwind of the Moscow vault

LETHALITY

Medium—30 percent for ordinary and modified forms of smallpox, almost 100 percent for flat and hemorrhagic smallpox

KILLS PER ANNUM

Zero . . . for the moment

HISTORIC DEATH TOLL

Hundreds of millions, if not over a billion. Historians believe that smallpox may have killed off a sizable population of the native North, South, and Central American population after Europeans with the disease carried it with them to the New World.

NOTABLE VICTIM

Janet Parker, a medical photographer at Britain's University of Birmingham Medical School, became the last ever victim of smallpox on September 11, 1978, when the virus made its way to her darkroom from a microbiology lab one floor below. After she died, the head of the microbiology department, Henry S. Bedson, committed suicide in despair.

HORROR FACTOR: 10

Smallpox ravages the entire body and even those who survive are often left disfigured and handicapped.

GRIM FACTS

- Samples of the deadly virus are stored in two vaults for future research: one in the Centers for Disease Control (CDC) in Atlanta, Georgia, and the other in the Research Institute for Viral Preparations in Moscow. There has long been debate about destroying the samples, and initial plans by the World Health Organization had called for their destruction before the year 2000. They're still there.

- Smallpox was named the "small" pox in order to distinguish it from the "great" pox, which was syphilis.

- Smallpox may have been the first biological weapon intentionally used for warfare. Sir Jeffrey Amherst, commander-in-chief of British forces in North America, recommended in 1763 that ground smallpox scabs be rubbed into blankets, which would then be given to American Indians.

- Smallpox is the only viral infection to ever be completely obliterated through scientific research.

SNAKES

Snakes are the antithesis of humans, a significant point in explaining why they are so psychologically frightening to both men and women. Snakes have no hands, no feet, no warmth, no brains. They slither on the ground and have fangs. And if you believe everything you read in the Bible, it was a snake that stole humanity's innocence and handed us our very first ticket to Hell.

Unlike other creatures that can and do kill people, the snake family is blessed with two radically different methods by which it can hasten your departure to the afterlife. One is via the liberal and painful administration of venom into your bloodstream. The other is by constriction, which literally squeezes the life out of you. This multipronged approach to killing entitles snakes to the only two-part segment in this entire book.

Making matters worse for you is that snakes don't like being bothered by humans . . . at all. Especially the big ones and the venomous ones. Every year, tens of thousands of people learn this lesson the hard way.

PART 1: VENOMOUS SNAKES

There are 600 venomous species of snakes, which is equal to one-fifth of all snake species. They are found on every continent except Antarctica, primarily in tropical and desert areas. And they live in holes, garages, basements, fields, trees, rivers, tunnels, and many other places where humans are likely to come upon them unexpectedly.

Before we begin, you need to know the difference between venom and poison. All venoms are poisons, but not all poisons are venoms. It's kind of like all hearses are black cars, but not all black cars are hearses. Venom is poison that is injected by animals into their prey, rather than a poison that is rubbed on, or eaten by, the victim. Snakes are regarded as venomous, while plants and frogs are considered poisonous. In the end, they'll both kill you, but it's good to know the technical difference, just for peace of mind.

HOW IT KILLS

Snakes strike quickly, typically at a rate of eight to ten feet per second. In this amount of time, it attacks, bites, and withdraws to observe the results of the bite.

Venomous snakes inject poison—a form of saliva, actually—into victims by puncturing the skin with two fangs. The venom is delivered into the puncture wound either via hollow fangs, like a hypodermic needle, or along a groove in the fangs. Depending on the species, fangs range in length from about a third of an inch to an inch.

Snake venom is a neurotoxic cocktail that attacks your body primarily by assailing the neurological system. Neurotoxins screw up your brain's control of your body by preventing communication between neurons. Essentially, your nervous system goes haywire, and the signal system that controls muscles and organs is shut down. The first symptoms you experience are headache, nausea, vomiting, abdominal pain, convulsions, and a loss of consciousness. Severe bleeding may result due to anticoagulants in the venom. Making it even worse, hemorrhaging can occur in the brain.

Then the fun really starts. Muscle pain transforms into neurotoxic paralysis. Proteins in the toxin begin to eat away at muscle tissue, causing it to dissolve. The urine turns reddish-brown as muscle tissue passes through the kidneys. The kidneys are damaged by filtering so much tissue debris out of the blood and acute renal failure follows. Eventually, the lungs are unable to inflate and death, mercifully, occurs.

KNOWN BY SCIENCE AS:

Taipan (*Oxyuranus microlepidotus*) found in Australia; Black Mamba (*Dendroapsis angusticeps*) found in Africa; the Common Krait (*Bungarus caeruleus*) found in Southeast Asia; Indian or spectacled Cobra (*Naja naja*) found in South Asia; the Diamondback Rattlesnake (*Crotalus atrox*) found in North America

MEDICAL CAUSE OF DEATH

Cerebral hemorrhage; respiratory paralysis

TIME TO KILL

Symptoms begin within half an hour after injection of venom; death may take several weeks to occur

HIGHEST RISK

Just about anyone living in Southeast Asia; hikers in the western United States and the Australian outback; fundamentalist preachers who think snake handling is something that impresses God

LETHALITY

Medium. Untreated snakebites have a lethality of 75 percent or more. Antivenom is readily available in the United States, so mortality is low, typically less than a dozen deaths per year. On the other hand, those getting bitten by taipans in Australia or cobras in India have little access to antivenom, and stand a good chance of becoming just as cold-blooded as their attackers.

KILLS PER ANNUM

In the United States, about a dozen people per year die after being bitten by venomous snakes (although nearly 8,000 are bitten). Worldwide, some 50,000 deaths are reported every year, although some researchers believe that there may be 70,000 more deaths that aren't reported. Most of the deaths occur in Southeast Asia.

HISTORIC DEATH TOLL

Millions. More than 50,000 people a year die from snake bites, so it takes just twenty years to rack up a cool million. And snakes have been ruining people's lives since Day One, according to the Bible.

NOTABLE VICTIM

Evangelist John Wayne "Punkin" Brown Jr. was bitten on the finger by a timber rattlesnake that he was waving around during a sermon at the Rocky House Holiness Church in Alabama, on October 3, 1998. His wife had died the same way three years earlier.

HORROR FACTOR: 7

It's not as bad as a lot of other animal attacks, and certainly preferable to dying by constriction. The attack and bite are done and over in less than three seconds from the moment the snake strikes. In fact, that part of it is over almost before you know it. The calculation you make about how much time you have to get to a hospital is probably the worst part of the experience, until your organs start shutting down.

GRIM FACTS

- Snake handlers believe they have control over venomous snakes because of the following verse from scripture: "And these signs shall follow them that believe: In my name shall they cast out devils; they shall speak with new tongues. They shall take up serpents; and if they drink any deadly thing, it shall not hurt them; they shall lay hands on the sick, and they shall recover" (Mark 16:17–18).

- Some cobras "spit" venom into their victim's eyes with incredible accuracy. However, it is more of a concentrated spray than a spit, so it will cover much of your face. While this won't kill you, it can cause permanent blindness.

- The black mamba, considered one of the world's deadliest snakes, can rear up to almost five feet off the ground while attacking. That's just about high enough to look you in the eye, isn't it?

PART 2: CONSTRICTORS

HOW IT KILLS

There are five species of constrictors big enough to kill humans: the reticulated python, Amethystine python, the Indian python or Burmese python, African rock python, and the Green Anaconda.

Because constrictors are so big—more than a dozen feet in length, weighing several hundred pounds—their ground speed is incredibly slow. Thus, they rely on ambush tactics, such as waiting for their prey in trees or in shallow water.

Unlike venomous snakes, constrictors do not have large fangs. Their teeth are relatively small and used for latching onto their victims. The teeth hold firmly while the snake's body loops around the prey in large muscular coils. This coiling serves an initial purpose of preventing escape, as well as to immobilize the victim so it doesn't fight back. Once this is achieved, the coils provide an excellent way to kill that victim—namely, you.

Every time you exhale, the snake tightens its coils to put pressure on your lungs and heart. Your lungs can't expand to get enough air, and your heart is unable to pump the required amount of blood to the rest of your body because blood vessels are squeezed shut. While scientists used to believe that prey suffocated from a lack of oxygen, it is now widely believed that most victims die from a heart attack or from not getting enough blood to their brains. In effect, circulation stops immediately, which is a more efficient and faster way to kill than through suffocation.

For all the squeezing and constricting, there is very little snapping of bones or breaking of your body. Constriction pressure is rather gentle, all things considered, as it only needs to prevent lung expansion and blood flow. Fortunately, death occurs before the snake unhinges its jaw and clamps it over your head before swallowing your entire body inch by inch.

KNOWN BY SCIENCE AS:

Amethystine python (*Morelia amethystine*); the Indian python or Burmese python (*Python molurus bivittatus*); African rock python (*Python sebae*); reticulated python (*Python reticulates*) and the Green Anaconda (*Eunectes murinus*)

MEDICAL CAUSE OF DEATH

Heart failure; occasionally respiratory failure

TIME TO KILL

From the moment a constrictor grabs you and starts coiling until the moment your heart gives up is less than half an hour

HIGHEST RISK

Amazon villagers; Southeast Asian farmers; idiotic pet owners who consider themselves "snake experts" when they buy baby constrictors

LETHALITY

Low. For all but the biggest snakes, like the anaconda, lethality is a function of pure chance. Since the constrictor is rarely in a hurry, there is time for a well-equipped human to escape death. This can be done by killing the snake before you lose consciousness or through the intervention of someone who finds you in this predicament. If the coils are already locked around you, however, we put your probability of death at over 90 percent.

KILLS PER ANNUM

Less than five

HISTORIC DEATH TOLL

Not high. Probably only several hundred in modern history.

NOTABLE VICTIM

Richard Barber, of Aurora, Colorado, was killed on February 10, 2002, when his eleven-foot-long python, named "Monty," strangled him to death. Barber had been told three years earlier to get rid of the snake by a judge who deemed he was in violation of keeping a snake longer than six feet long in the city.

HORROR FACTOR: 10

Unlike the swiftness inherent in a venomous snake attack, the duration of death by a constrictor is going to feel like an all-day pig roast in comparison.

GRIM FACTS

- Pythons and anacondas can grow to over thirty feet long, and anacondas can weigh more than half a ton.

- If an anaconda digests all of you, it may not need to eat for an entire year.

SPACE SUIT MALFUNCTION

When you leave Earth's atmosphere, you damn well better make sure that you have a pressurized space suit. If you don't, you'll end up boiling mad . . . and boiling to death.

HOW IT KILLS

At 250 miles above the surface of the Earth, about the altitude at which the space shuttle flies, space is a huge vacuum devoid of gas. There is no air to breathe, which is why astronauts are equipped with oxygen in their space suits. These space suits protect astronauts from something even more lethal—the absence of atmospheric pressure. With no external pressure on the human body, internal processes very desperately agitate to become external processes.

Imagine you're on a space walk, repairing one of those pesky space station arms that always seems to keep breaking down. You maneuver outside the cabin and all of a sudden your pressurized space suit rips right open. You might be able to hold your breath for a few minutes, but that's the least of your worries. With no pressure from your suit, the inside of your body completely decompresses, attempting to equalize with the vacuum in which you're floating. All the air in your lungs will be sucked out, which wreaks havoc on your lungs. The oxygen and nitrogen that is traveling along in your bloodstream, which had dissolved under pressure, starts expanding, looking for a means to escape from your blood. In effect, your entire body is like a human soda can—all shaken up—just waiting to be opened.

We all know what happens when that can opens. The gas bubbles rise up and froth over, which is what will occur inside you. With no pressure on the gas contained in the fluids in your body, your blood begins to boil. Yes, boil. Reducing pressure on a gas causes it to boil in the same way that heat does. Also, the liquid in your eyes, tongue, nostrils, bladder, and everywhere else boils and turns to vapor. As the vapor tries to escape from its surroundings, the liquids and solids in your body separate, rupturing your blood vessels and ripping apart your tissue and organs. The expanding gas causes your body to swell to twice its size. Fortunately for you, it only takes fifteen seconds of this to render you unconscious, because you don't want to be awake through the rest.

In the next three to four minutes, your brain is completely emptied of oxygen and shuts down for good. Gases escape from every orifice they can and are sucked out into the void. After five minutes, all that's left of you is a completely dried-out husk of a human.

KNOWN BY SCIENCE AS:

The condition of being suddenly exposed to a vacuum due to an equipment malfunction is called "explosive decompression." *Ebullism* is the term for the formation of bubbles in your bodily fluids because of an extreme reduction in the surrounding pressure. The reason for your boiling blood is explained by Henry's Law, which states that "the form of the equilibrium constant shows that the concentration of a solute gas in a solution is directly proportional to the partial pressure of that gas above the solution."

MEDICAL CAUSE OF DEATH

Asphyxiation; ebullism; brain damage

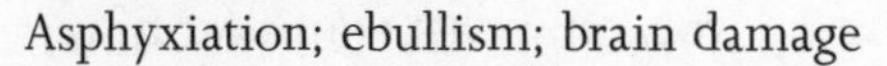

TIME TO KILL

Five minutes, tops

HIGHEST RISK

Astronauts; test pilots; future space travelers

LETHALITY

High—100 percent. Unless you have a sewing kit to repair that suit—and can fix it in less than fifteen seconds—you are going to turn to froth. However, if you are close to a decompression chamber and someone can get you into it within two minutes, there is a small chance that you will survive.

KILLS PER ANNUM

Less than one

HISTORIC DEATH TOLL

Confirmed at three (see below), although some early high-altitude test pilots may have died from inadequate pressurization

NOTABLE VICTIMS

On June 30, 1971, the *Soyuz II* spacecraft lost all pressure due to a malfunction just before re-entry. The three Russian cosmonauts, Georgi Dobrovolski, Vladislav Volkov, and Viktor Patsayev, were not wearing space suits at the time of the accident, and died in forty seconds.

HORROR FACTOR: 10+

The moment you hear that suit rip, you know that your next fifteen seconds are going to be about as crappy a time period as you've ever had in your life.

GRIM FACTS

- In 1966, a NASA (National Aeronautics and Space Administration) test subject was accidentally exposed to a near vacuum when his space suit developed a leak in a pressure chamber. He reported feeling the water in his mouth boil before passing out after fourteen seconds. He did not die.

- A former NASA employee claims that an astronaut on STS-37 (the eighth flight of the space shuttle *Atlantis*) suffered a punctured a glove during a spacewalk in April 1991, but the tiny hole was immediately plugged by the astronaut's blood.

- Many scientists, including some within NASA, refuse to acknowledge that blood or other bodily fluids "boil" in a vacuum, apparently in the belief that "boiling" connotes the use of heat. They instead describe ebullism with the more passive phrase "liquids turn to vapor."

STAMPEDING AND TRAMPLING (HERD ANIMALS)

Plenty of animals kill you with their teeth, their claws, their venom, and their strength. There are a lot of animals that will kill you just by running right over the top of you.

HOW IT KILLS

Herd animals, notably elephants and bovines such as buffalo and domestic cattle, have been known to trample humans during a stampede. Individual cases of trampling are rare, although they have been known to occur when a large animal has been antagonized by a human and fights back. This has been frequently reported in circuses and on farms.

Stampedes are not fully understood but are typically believed to be reactions to a single frightening incident that triggers the flight instinct in mammals. The result is a kind of running mass hysteria, and little other than exhaustion or an immovable obstacle can stop a stampede.

Trampling involves the animal in question placing its weight on various parts of your body, repeatedly. The animal—or multiple animals in the case of a stampede—finds that you are in its path and has decided it is not going to let you stop it from getting to wherever it wants to go. Animals like wild horses, buffalo, and elephants, which weigh from a quarter ton to several tons, simply mow over you. Depending on the point at which their leg, and thus a significant proportion of their weight, comes down on you, severe trauma will result. Stepping on your head is likely to mean instant death, while the pressure applied to other organs will cause immediate organ failure, especially if lungs are collapsed or the heart is crushed.

KNOWN BY SCIENCE AS:
Trampling

MEDICAL CAUSE OF DEATH
Organ failure, compressive asphyxia

TIME TO KILL
About three minutes should do it. If a herd tramples you, best to hope it's a big one so that it gets over with quickly.

HIGHEST RISK
Cowboys; the slowest sprinters among those running with the bulls in Pamplona; cocky matadors; villagers who live near thirsty elephants

LETHALITY
Very high. The weight of a large animal, as well as the force with which it uses that weight, results in little time to get away or fend off the attack.

KILLS PER ANNUM

A dozen. These are often cruel animal trainers, overly curious safari-goers, or the unlucky victims of a farming accident.

HISTORIC DEATH TOLL

Unknown, but when accounting for the use of horses and elephants in ancient warfare, the estimate is several thousand.

NOTABLE VICTIMS

On July 10, 1972, a severe summer drought in Bhubaneshwar, India, reached the boiling point when thirsty elephants descended on the tiny village. The crazed beasts roared out of the nearby Chandka Forest and trampled the community and its dwellings, killing twenty-four people.

While walking her two dogs along a footpath in Suffolk, England, forty-five-year-old Sandra Pearce was trampled to death by a herd of stampeding cows. The April 28, 2008, incident was thought to have been prompted by the barking of Pearce's terriers.

HORROR FACTOR: 8

There's no way to get out of the way of a spooked or angry elephant or enraged buffalo, and this knowledge is intensified by the fact that you can actually see the big animal bearing down on you.

GRIM FACTS

- Prior to the 1800s, criminals and political dissidents in India were frequently executed by having elephants trample them. Specially trained elephants were also able to pull the limbs off of the condemned.

- Human stampedes, which occur at concerts, religious festivals, and sporting events, kill many more people per occurrence than animal stampedes. In human stampedes, asphyxiation accompanies organ failure as a common cause of death.

Heading for the Finish

Driving his beloved race car "Babs," British speed racer John Godfrey Parry-Thomas attempted to set a land speed record at Pendine, Wales, on March 3, 1927. However, during the run, Babs experienced severe mechanical trouble and her drive chain snapped, partially decapitating and killing Parry-Thomas. His body was sent back to England for burial while Babs was buried in the sand where the event occurred.

STROKE

Also known as a "brain attack," stroke is the Number 3 cause of death in the United States, and reigns as Number 1 in causes of disability.

HOW IT KILLS

A stroke is to the brain what cardiac arrest is to the heart. It occurs when the brain is cut off from blood flow, which provides your gray matter with the oxygen and nutrients it needs to keep you blinking, breathing, and saying clever things at parties. Without blood flow, brain cells die, and you don't have to be a neurosurgeon to understand that's kind of a bad thing.

There are two main types of stroke, and you don't want either one. Ischemic stroke, which accounts for 85 percent of all strokes, is when arteries are blocked by clots or by the buildup of plaque and other fatty

deposits. Hemorrhagic stroke is when a blood vessel leaks or breaks, releasing blood into the brain (*see* **Brain Aneurysm**). Each of these conditions kills brain cells, which die at a rate of two million per minute. Granted, not all of the cells in your brain are hard at work, but most of them are keepers.

When stroke strikes, it's usually without warning. You may first experience sudden numbness or paralysis on one side of your body, though it's not as if you could play a piano concerto with the "good" side. Since the functions of both body and mind are governed by different areas of the brain, your symptoms will vary depending on which side of the brain is experiencing cell death. Injury to the left side of your brain will lead to weakness or paralysis to the right side of your body. A lefty stroke will also leave you speaking in slurred tones, with your logic impaired—you'll mix up yes and no, and have trouble naming even simple objects. Injury to the right side of your brain leads to paralysis on the left side of your body, and will affect functions like recognizing faces, making decisions, or getting dressed. Over time, the complications resulting from a stroke on either side become more severe and increasingly disabling, until you can hardly keep the drool in your own mouth and even small children can beat you at checkers.

Of course, those are only problems if you survive the initial attack, so look at the bright side: On average, two out of every five strokes is a killer.

KNOWN BY SCIENCE AS:

Cerebral infarction

MEDICAL CAUSE OF DEATH

Cerebral ischemia; cerebral hemmorhage

TIME TO KILL

Minutes, hours, or months, depending on severity

HIGHEST RISK

African American males over sixty-five with diabetes and a family history of stroke—no other disease shows greater disparity in mortality rates between African Americans and Caucasians

LETHALITY

High. Roughly 40 percent not including subsequent deaths from stroke-induced complications

KILLS PER ANNUM

Approximately 144,000 in the United States but globally may exceed 5 million

HISTORIC DEATH TOLL

Estimated at several hundred million. However, improved control of risk factors—including hypertension, diabetes, bad cholesterol, and the use of some birth control pills—has decreased incidence. Death from stroke was 4 times more likely in the 1950s than it is today.

NOTABLE VICTIM

Richard Knerr of Wham-O toys had many strokes of genius including the Hula Hoop, the SuperBall, the Frisbee, and Silly String. His last stroke, in January 2008, was of the fatal kind.

HORROR FACTOR: 7

Strikes like a bolt of lightning to your brain

GRIM FACTS

- The layman's term "mini-stroke" alludes to a transient ischemic attack (TIA), which is a temporary interruption of blood flow to the brain. While the symptoms of TIA are similar to stroke, it doesn't kill brain cells or cause permanent disability.

- In 2004, the National Stroke Association developed the "Hip Hop Stroke" campaign to educate urban kids about the condition through fun activities like "Make a Jello Brain."

- Health experts say that 500,000 of the nation's 780,000 strokes every year are preventable, brought on by lifestyle choices (smoking, alcohol, poor diet, lack of activity) or limited access to proper health care for treating blood pressure and heart conditions.

SYPHILIS

Some sexually transmitted diseases make you itch. Some make it burn when you pee. Some eat away at your body until your bones are hollow and your brain descends into dementia. That last one would be syphilis.

HOW IT KILLS

Syphilis has been recognized for so long, and is so easily cured, that it seems like this STD would have been wiped out ages ago. But incidence has risen steadily since the year 2000.

The bugs at work in a case of syphilis are spirochetes, a corkscrew-shaped bacterium (think fusilli pasta) that typically infect broken skin or the membrane around the genitalia. Transmission is predominantly via sexual encounter, though it can also be communicated by blood transfusion, by kissing someone with an active lesion around the mouth, or from an infected mother to her unborn child. Nearly half the babies, born or unborn, who contract the disease from a syphilitic mother will die.

The rest of us stand an excellent chance of surviving if the disease is diagnosed within the first year, since primary syphilis (up to three months after exposure) can be cured with penicillin. But, left untreated, the spirochetes eventually accumulate in the body, causing lesions called "gummas" in your bones, skin, nervous tissue, heart, and arteries. The way syphilis kills depends on which tissue has been infiltrated and eaten away. Lesions in your heart can cause aneurysms and heart disease. Around the brain and spinal cord, syphilis leads to stroke or deadly meningitis. Other invasions of the central nervous system, known as neurosyphilis, can cause incontinence, blindness, paralysis, and dementia before death sets you free.

KNOWN BY SCIENCE AS:

Primary, secondary, or tertiary (late) syphilis, caused by the spirochete *Treponema pallidum*

MEDICAL CAUSE OF DEATH

Destruction of tissue leading to organ or system failure

TIME TO KILL

Without treatment, late-stage syphilis can onset one year after infection, though it's more typical around eight to ten years. Once at the advanced tertiary stage, all bets are off.

HIGHEST RISK

Men who have sex with men; sexually active women aged twenty to twenty-four; and men aged thirty-five to thirty-nine

LETHALITY

Low. More than 32,000 cases of syphilis were reported to the Centers for Disease Control (CDC) in 2002; approximately one in 1,000 will die of the disease. (Compare that to 89,000 syphilis deaths in Africa in 2002.) Nearly half the babies, born or unborn, who contract the disease from a syphilitic mother will die.

KILLS PER ANNUM

More than thirty in the United States

HISTORIC DEATH TOLL

Tens of millions, if not hundreds of millions. An outbreak in Europe during the late 1400s, which may have killed ten million alone, is widely believed to have been brought back to Spain by Columbus after visiting the New World.

NOTABLE VICTIMS

Syphilis has been identified or suspected in the deaths of a disproportionate number of artistically minded people through the ages, including Charles Baudelaire, Beau Brummell, Édouard Manet, Paul Gauguin, Scott Joplin, Al Jolson, and Howard Hughes. It also seems to go after historic bad-asses like Ivan the Terrible, Frederich Nietzsche, Vladimir Lenin, Al Capone, and Idi Amin.

HORROR FACTOR: 5

Way worse than a case of crabs, though not as bad as being mauled and pincered to death by actual giant crabs

GRIM FACTS

- According to 2002 statistics by the American Social Health Association (ASHA), the rate of syphilis among African Americans was eight times higher than for whites.

- Syphilis gets its name from the poem "Syphilis sive morbus gallicus," written in 1530, by the Italian Girolamo Fracastoro. The title translates as "Syphilis or The French Disease"—though the French at the time called it "The Italian Disease."

- Of the 32,000 U.S. cases reported in 2002, nearly half came from sixteen counties in the Southeast.

- In the ill-famed Tuskegee Syphilis Experiment, public health officials conducted a forty-year study on 399 black men, most of them poor sharecroppers from Alabama. The trusting men thought they were receiving medical care, but the "experiment" amounted to a long-term study of how long it would take them to die from syphilis.

- Genetic evidence suggests that Christopher Columbus brought syphilis to Europe when he returned from the New World—leading directly to the European syphilis epidemic of 1495.

TETANUS

Your mother always warned you that if you got hurt by stepping on a rusty nail, you'd have to get a tetanus shot—which seemed scary enough. The shot is nothing compared to the horror that is the lockjaw of tetanus.

HOW IT KILLS

Tetanus is created by the anaerobic bacterium *Clostridium tetani,* which lives as dormant spores in the ground (much like anthrax). The spores enter your body through an open wound, such as when you step on a dirty nail, get stabbed by a rusty knife, or are cut by a gardening utensil. Once firmly situated in the low-oxygen confines below your flesh, the spores germinate and the bacteria starts to spread.

It releases a neurotoxin called tetanospasmin, which on the basis of weight is one of the most toxic substances anywhere. It takes 175 nanograms (a nanogram is one billionth of a gram) to kill an average adult. The toxin is spread away from the wound by the bloodstream, where it makes its way to the brain and spinal cord. There, it blocks the "off" signals the brain sends to muscles.

The first sign that you have tetanus is that your jaw will tighten, and you'll find it difficult to open your mouth. This is known as "lockjaw," and the effect is similar to that of having your jaw wired shut. Your

neck will also stiffen, and your vocal cords will start to spasm, making it hard to swallow. Eating solid foods will become nearly impossible, and even if you felt like you could lose a few pounds, this isn't the way to go about it.

The stiffness will extend down your body and you'll feel a tightening around your abdomen. Your chest muscles and your throat will spasm, restricting your ability to breathe. Fever and profuse sweating add to the discomfort.

Your entire body will start to spasm as your muscles contract against your will. The long bones in your body, such as those in your arms and legs, will fracture as they get snapped around by your out-of-control muscular system, and your muscles will actually tear from the strain. As your back arches from being pulled tight by muscles, your spine might crack.

The wild swings of your autonomic nervous system may also affect your heart, leading to arrythmia. The inability to control your respiratory muscles will most likely lead to asphyxiation. Interestingly, tetanus is one of those diseases that does not always have a specific cause of death other than the effects of the neurotoxin.

KNOWN BY SCIENCE AS:

Tetanus is caused by the *Clostridium tetani* bacterium

MEDICAL CAUSE OF DEATH

Asphyxiation; heart failure; direct effects of tetanospasmin on the nervous system

TIME TO KILL

Symptoms show up after a week, and death occurs approximately three to four weeks later

HIGHEST RISK

Infants born in rural areas without hospitals; bare-footed farm workers

LETHALITY

Low, about 10 percent. Tetanus shots are available once infection has begun, and most people in developed nations have been immunized.

KILLS PER ANNUM

Approximately 300,000, primarily from neonatal tetanus, which kills babies who are delivered using unsterilized or dirty surgical instruments

HISTORIC DEATH TOLL

Millions upon millions

NOTABLE VICTIM

John Augustus Roebling, the designer of the Brooklyn Bridge, contracted tetanus after his toes were crushed by a ferry while he was standing on a dock mapping out the bridge's location. He died twenty-four days later on July 22, 1869.

HORROR FACTOR: 8

The first signs of lockjaw are only an indication of how completely out-of-control your body is going to be.

GRIM FACTS

- Three boys from New Jersey died from lockjaw at Christ Hospital in Jersey City Heights on July 12, 1908. They had all been injured on different days in different parts of the state, yet each one was admitted for injuries related to gunfire.

- There has long been a popular notion that rusty nails cause tetanus. In fact, any sharp object that introduces the *Clostridium tetani* spores deep into a wound can cause tetanus. That includes glass, human teeth, metal tools, and wood splinters. A sterile nail will not cause tetanus.

TRYING TO BEAT A LOCOMOTIVE ACROSS THE TRACKS

Is that the white light at the end of the tunnel or . . .

HOW IT KILLS

Trains generally take about one mile to stop. That rule of thumb holds true for a mile-long, 100-car freight train moving 55 miles per hour and for an eight-car passenger train speeding along at 80 mph. Considering that each freight car can weigh over 140 tons, with dozens of cars propelled by the force of a 7,500-horsepower locomotive, it's not hard to see that the laws of physics are working against someone bending over to pick up a nickel. You wouldn't even stand a prayer crossing the tracks in a two-ton pickup truck.

In a train-versus-automobile crash, the passenger vehicle explodes into an unrecognizable heap of twisted metal. Imagine what it does to the soft mass of skin and bones riding inside.

Hapless drivers who knowingly try to beat a train across a grade crossing are often victims of their own miscalculation. Much as an airplane appears to be moving slowly in the sky, the speed and distance of an oncoming train is easily misjudged.

Thousands of the nation's 250,000 level crossings have no warning lights or gates to caution drivers who would very much like to make it to the other side of the tracks alive. It may seem unlikely that something as huge and loud as a locomotive could take a driver by surprise, but anyone who has inched a vehicle past overgrown vegetation to peer down the tracks knows better. Plus, some drivers forget to check for a northbound train once the southbounder has passed safely by.

Back in the 1990s, the National Highway Traffic Administration prepared a study on rail crossings and offered a profile of motorists who die in crossing crashes. The most likely victim, their statistics revealed, is a white male driving in Texas, who reads *Field and Stream*, listens to country music, and chews tobacco. This doesn't mean that you can't catch steelheads and enjoy a chaw without getting hit by an iron horse, but you might want to turn down the Conway Twitty at a crossing.

KNOWN BY SCIENCE AS:
Highway-rail crossing incident

MEDICAL CAUSE OF DEATH
Blunt force trauma resulting in system failure

TIME TO KILL
Instantaneous, or prolonged following injuries

HIGHEST RISK
Rural drivers at unmarked crossings

LETHALITY

Medium—12 percent of all crossing incidents result in fatality

KILLS PER ANNUM

The current annual average (as calculated using five years of fatality stats provided by the Federal Railroad Administration) is 355—very close to one death every day in the United States.

HISTORIC DEATH TOLL

Thousands. Between 2000 and 2004, more than twice as many Americans died at crossings than in commercial airplane crashes.

NOTABLE (ALMOST) VICTIM

Text messaging while driving a vehicle is an increasing hazard among teenagers, and texting while crossing train tracks definitely ups the ante. In 2007, an eighteen-year-old in Ohio nearly cashed all his chips when he was sending an SMS from his phone while crossing tracks on foot. He was struck by a train and thrown fifty feet, but lived to text again. OMFG.

HORROR FACTOR: 4

The terror is paralyzing but mercifully brief.

GRIM FACTS

- A 2004 investigation by *The New York Times* revealed that some railroads skirt crossing-maintenance responsibilities and underreport fatalities.

- The worst crossing crash occurred in July 1967, in Langenweddingen, Germany (then East Germany), when a barrier malfunctioned.

A train hit a lorry which burst into flames, killing 94 people including 44 children on vacation.

- In the 1986 film *Stand By Me*, based on a Stephen King story, four boys seek the body of young Ray Brower, who had been hit by a passing train.

TUBERCULOSIS

One of history's most effective and long-lasting killers, tuberculosis is like a really, really bad cold, until you start coughing up blood.

HOW IT KILLS

Tuberculosis is caused by a common form of tube-shaped bacteria known as *Mycobacterium tuberculosis.* It is spread, like pneumonia, via infected water droplets that you inhale when an infected person breathes, talks, spits, or otherwise propels those water droplets your way. You should note that one in three people probably have a latent form of the disease, but only 10 percent of those have the active form. Stay the hell away from them. Because if you get the active form, TB gets grim quickly and doesn't let go for weeks.

Initially, you will have chest pains, and begin to cough. That cough will persist and get stronger, until you start coughing up blood. You'll start losing weight, and experience both fever and chills, while losing even more bodily fluids through ongoing bouts of the night sweats that soak your sheets like buckets of water. You'll also be incredibly tired much of the time you're awake. Your skin will

become a deathly white, and you may develop a sensitivity to bright light—giving your daily life a somewhat vampiric tone when coupled with the blood dribbling out of your mouth from all the coughing.

Liquid will fill up the space around your lungs, making it difficult to breathe. The bacteria then move to your lymph glands and to other organs and tissues, notably your spine and the meninges of the brain. When the infection attacks your spine, it prevents nutrients from getting to your vertebrae, which then start to corrode and damage the support structure of your spinal column. This will twist your spine into a curve, if you don't die first. Then tuberculosis proceeds to your bone marrow, infecting your blood at its source. When the disease hits your bladder, you will start pissing blood, and from there it will go on to destroy your reproductive organs. When your entire body can't take any more, the show will be over.

KNOWN BY SCIENCE AS:

The bacterium most commonly responsible for tuberculosis is the *Mycobacterium tuberculosis*. It is also known as the *tubercle bacillus*, which is where the initials TB come from. The dangerous form of tuberculosis currently extant is called multidrug-resistant TB.

MEDICAL CAUSE OF DEATH

Tuberculosis is considered its own cause of death because of its attack on so many different parts of the body. Respiratory failure and organ failure are two common end results of TB infection.

TIME TO KILL

Approximately three weeks for a severe case; years and even decades for a prolonged illness

HIGHEST RISK

Kids in developing countries; AIDS patients; alcoholics; people living in crowded conditions

LETHALITY

High. Currently over 50 percent and getting higher. Tuberculosis has become resistant to many forms of medication in the last two decades, and the susceptibility of AIDS patients to the disease has raised the number of cases.

KILLS PER ANNUM

Somewhere between two and three million, but probably a lot more

HISTORIC DEATH TOLL

Billions. Prehistoric humans had TB, and no way to treat it. TB is estimated to have killed one out of every four people in Europe during the seventeenth and eighteenth centuries, and another 100 million in the twentieth century alone. It is not called "the captain of the men of death" for nothing.

NOTABLE VICTIMS

For some reason, writers and musicians have been hit hard by tuberculosis over the years. Among the most famous to die from the disease are Franz Kafka, George Orwell, D. H. Lawrence, Anton Chekov, and Frederic Chopin. Other famous deaths include Vivien Leigh, President Andrew Jackson, Eleanor Roosevelt, and "Doc" Holliday.

HORROR FACTOR: 3

It's a bad cough that you have for a long time, and it could be for years. Coughing up blood is no fun, however, and neither is that gradual curvature of your spine.

GRIM FACTS

- Worldwide, one person is infected with TB every single second of every day.

- At the turn of the twentieth century, TB was the leading cause of death in the United States. At the time it was referred to as consumption, leading to this little ditty in the pre-antibiotic days: "TB or not TB, that is the question. Consumption be done about it? Of cough not."

- TB is a major cause of death among people who have AIDS.

- The first antibiotic to fight TB, Streptomycin, was introduced in 1946. On December 4, 1967, U.S. Surgeon General William H. Stewart declared that infectious diseases like tuberculosis had been defeated—a premature announcement at best. In 1993, the World Health Organization (WHO) declared tuberculosis a global emergency.

UNSCHEDULED PLANE LANDING

Evolution has not prepared us to be shot through the air in a metal tube traveling 600 miles per hour. An unscheduled landing anywhere but on an airstrip reminds us why.

HOW IT KILLS

As proponents of air travel often note, you are statistically more likely to survive a plane trip than a ride to the airport. The chances of being involved in an aviation accident are, for the typical traveler, about one in 11 million, compared to a one in 5,000 chance of being killed in a car wreck. Psychologists suggest it is the sheer loss of control that induces panic; you might be able to steer a car out of a deadly skid, but in an airplane all you can do is place the tray in an upright position and pray that the pilot is sober.

A turbulent descent leaves passengers white-knuckled and clutching family pictures. However, accidents during final approaches and landings actually account for fewer fatalities (18 percent) than the initial climb, that flaps-up phase of flight during which the greatest percentage (25 percent) of deadly disasters occur.

Those with a fear of flying should take a Dramamine now, and skip the following paragraph.

Pilot error has long led the list of possible causes for a plane landing in someone's yard instead of at an airport. Others include air traffic control error; bird strike; fire in the cabin or cargo hold; aircraft design flaw; maintenance and repair negligence; sabotage and hijacking; out of gas (known in the industry as "fuel starvation"); lightning; and pilot incapacitation.

KNOWN BY SCIENCE AS

Aviation accident

MEDICAL CAUSE OF DEATH

Blunt force trauma resulting in system failure; carbon-monoxide poisoning (smoke inhalation); incineration

TIME TO KILL

Usually instantaneous

HIGHEST RISK

Nonscheduled "on demand" commuter flights are twice as likely to result in fatality than a scheduled commuter flight, and thirteen times riskier than a traditional passenger airliner

LETHALITY ☠☠☠

High. However, the odds of being killed on a flight, provided the airline is reputable, is one in 10.46 million. Jump on an island-hopper piloted by a guy wearing sandals and the odds leap to one in about 725,000.

KILLS PER ANNUM

In a good year, approximately 1,000 worldwide

HISTORIC DEATH TOLL

Numbers are incomplete, due to poor recordkeeping in various countries, but total deaths are estimated at just over 100,000 (including wartime aviation deaths) since the start of regular commercial service in the 1920s. 2007 was the safest year for air travel since 1964. The worst year was 1972, when disasters claimed 3,214 lives.

NOTABLE VICTIMS

When Uruguayan Air Force Flight 571 crashed into the Andes in October 1972, survivors resorted to cannibalism, taking the idea of bad airplane food to an all-time low.

HORROR FACTOR: 8

Barring the possibility of surviving only to be incinerated by ignited jet fuel, the horror of a crashing plane is not pain on impact—it's in knowing, on the physically painless descent, that the end is nigh.

GRIM FACTS

- The accident with the most fatalities on record occurred in March 1977, when a KLM 747 collided with a Pan Am 747 while the Pan Am plane was taxiing on a runway in the "Fortunate Isles" (aka Canary Islands); 583 of the 644 people aboard both planes perished.

- More than 180 air carriers worldwide have never logged a fatality.

Guess Who's Coming to Dinner?

Bernd Jürgen Brandes, a forty-three-year-old engineer in Germany, responded to an advertisement by hopeful cannibal Armin Meiwes, asking for a willing victim to "be slaughtered and then consumed." Brandes apparently found this an exciting way to die, and he and Meiwes met on Christmas Day, 2001, to begin the process. The event was captured on videotape, which began with the amputation and eating of Brandes's penis, resulting in massive blood loss, proceeded on to consumption of pills and alcohol, and ended with Brandes' being stabbed in the neck by Meiwes.

WORKING IN A COAL MINE

Every foot below the surface gets a coal mine worker a little closer to Hell. When things go wrong, it's far worse than being just six feet under.

HOW IT KILLS

Coal, which generates 40 percent of the world's electricity, can be extracted by blasting away mountaintops and surface mining in shallow reserves. To reach deeper coal sources, we send men underground.

It's a long way down. Miners descend as far as 1,000 feet below the surface—the equivalent of 100 stories—in an elevator before the real digging begins. Once at depth, they head sidelong into the earth for distances that can exceed 2.5 miles. As you can imagine, it's a long way out when trouble strikes, and it often comes in the form of a controversial practice known as "retreat mining." That's when miners attempt a controlled roof collapse as they work their way back out of a mine. Because the pillars used to support a roof are themselves made of coal, the companies want their people to recover the pillars—regardless of the fact that retreat mining accounts for a third of all coal-mining disasters. When miners pull the pillars, the collapsing mountain of earth above sometimes leaves them trapped or crushed.

Mine blasts are another primary cause of cave-ins and historic disasters. If you've ever seen sparks fly from hitting a rock with a shovel, you can understand the risk. Naturally occurring flammable gases, known collectively as firedamp, can ignite when a miner's metal tools throw sparks. The ignited gasses in turn ignite coal dust and then—*boom*—there's an underground explosion. Such a blast was responsible for the worst coal catastrophe in history, when 1,549 workers died on April 26, 1942, in China's Liaoning region.

If you are not crushed or burned in a collapse, then you are in a race against the clock to access breathable air. If you have some air, perhaps from a small underground pocket, your best bet is to attempt to signal your location by banging on mine bolts or conveyor rails. Help has to arrive fast, because the air doesn't last long. In addition, inhaling smoke or noxious fumes (such as methane) will eventually cause you to lose consciousness . . . unless you can be dug out or if air shafts can be tunneled to your location. Don't hold your breath waiting for that to happen.

KNOWN BY SCIENCE AS:
Occupational hazard

MEDICAL CAUSE OF DEATH
Asphyxiation; fatal crush injury

TIME TO KILL
Seconds in the case of a blast or flood; days in the case of asphyxiation

HIGHEST RISK

West Virginian workers in deep-mining operations—West Virginians account for more than 25 percent of all coal fatalities since 1996

LETHALITY

Medium. Still considered among the most dangerous of all professions. The fatality rate for full-time coal workers is just under thirty per 100,000 full-time employees

KILLS PER ANNUM

Fewer than 50 in the United States compared to an average of 5,000 every year in China. U.S. fatalities had reached an all-time low of twenty-two in 2004, but spiked again in 2006 following the Sago Mine disaster in West Virginia (12 dead) and another in Kentucky (5 dead).

HISTORIC DEATH TOLL

Hundreds of thousands. France and China both had disasters that each claimed more than 1,000 lives. In 1907 alone, the United States recorded 3,242 deaths, including a disaster in Monongah, West Virginia, that claimed 362 victims.

NOTABLE VICTIMS

Chinese coal workers are now dying in numbers unprecedented even in the worst U.S. years. Approximately 3,800 Chinese were killed in 2007, and that number was down 20 percent from the year before.

HORROR FACTOR: 9

It's dark, you're underground, you have no food or water, it's unlikely anyone will get to you—and you're running out of air

GRIM FACTS

- It doesn't take a collapse or explosion to do in a coal miner. Fumes from leaking methane or other naturally occurring gases can be enough to asphyxiate underground workers.

- Coal workers' pneumoconiosis, also known as black lung disease, is still responsible for the deaths of over 850 miners every year.

- The expression "canary in a coal mine" originates from an antiquated practice of carbon monoxide detection. The caged bird would demonstrate signs of distress—such as swaying or falling from a perch—when it encountered the deadly gas.

THIS IS THE END

AFTERWORD

BY BILL MCGUINNESS, FUNERAL DIRECTOR

MCGUINNESS FUNERAL HOMES

WOODBURY, NJ

From the vantage point I hold as a funeral director, many people think I have a special insight into all matters death. I'm not certain I do, but after twenty-five years in this profession, I do have some observations.

The preeminent observation is that death is overrated. It's the finality of death that still gets to me. No one gets to say, "I made a mistake. How about a do-over?" A lifetime of learning, doing, achieving is gone—poof, outta here, thanks for stopping by. I guess that's why so much emphasis is placed on teaching our children, so they'll teach their children and those lessons will be passed on to the next generation.

That, of course, is the hook in life. Much of what we do ends up being an integral part of the way we die. Over the years, I've seen the result of virtually every calamity known to man. Car accidents by the dozens, drug overdoses by the scores, and a surprising number of people in their forties who die from scuba diving, swimming accidents, glider crashes, and dirt bike mishaps. Sometimes, I think, these activities are just "too much fun."

Then there are the more horrific deaths: skyscraper accidents, airplane crashes, infanticide, matricide, patricide. If it's been on the six o'clock news, I've seen it. Too many times.

I have come to the conclusion it's better to keep living. Better to keep working, laughing, and playing. Whatever it is that makes you unique, you should keep doing it.

Certainly, one can live too long. However, the longer one lives the less tragic the death is likely to be. That's the nature of how we deal with the death of the young and the old. And we all must deal with it. Many years ago, a friend's sister was killed in a scuba-diving accident. At the funeral, my friend said, "It's easy to ask 'Why us?' But the reality is 'Why not us?' We're not any different than any other family who meets tragedy face to face." Every one of us will come face to face with dying and death.

There is one truism about death that I call my own. I've found that it is extremely unlikely that you will die before the age of twenty-five if you don't have a tattoo. I first noticed a correlation between tattoos and young death many years ago, even before tattoos became wildly popular. Whenever someone young died, I would ask my staff if the person had a tattoo. Initially, the answer was always "Yes." Over time, however, the answer to the question became "Of course."

On those rare occasions when the victim doesn't have a tattoo, my response to the answer is always an astonished "Really?" I find that such an exception to the rule makes me more inquisitive about what happened. It makes me curious about the ways in which we die, and the things that kill us.

It is our humanity, our need to dare and to explore, that often brings us in close contact with elements that can destroy us. But if we didn't test our limits, we wouldn't grow.

There is a fine line between testing our limits and being reckless. I see the result when that line is crossed.

SOURCES

The statistics and scientific information in *This Will Kill You* were obtained from a variety of sources. These ranged from physician interviews and medical journals on to white papers, surveys, and the publications of numerous accredited agencies.

In cases where there was conflicting data among sources, we based our information on those with the most supporting documentation. When no specific statistics were extant, we based our estimates on extrapolation of relevant data.

Several hundred sources were used for the research in *This Will Kill You*, including the following:

AGENCIES AND ORGANIZATIONS

The Centers for Disease Control and Prevention; World Health Organization; National Institutes of Health; Federal Bureau of Investigation; American Heart Association; National Cancer Institute; National Library of Medicine; Federal Railroad Administration; Agency for Toxic Substances and Disease Registry; National Institute of Allergy and Infectious Diseases; American Medical Association; American Academy of Allergy Asthma & Immunology; Men's Health Network; United States Navy; NASA; Alzheimer's Foundation of America; National Organization for Rare Disorders; Mine Safety and Health

Administration; National Highway Traffic Safety Administration; Centre for Research on the Epidemiology of Disasters; National Fire Protection Association; Death Penalty Information Center; United States Department of Agriculture's Agricultural Research Service (ARS); Earth Institute at Columbia University; National Institute on Alcohol Abuse and Alcoholism; United States Department of Justice; United States Census Bureau; and the United Nations.

HOSPITALS AND LABS

Massachusetts General Hospital; Mayo Clinic; Cleveland Clinic; The University of Virginia Health System; University of Maryland Medical Center; UCLA Department of Epidemiology; New York University Medical Center; Weill Cornell Medical Center; and Columbia University Medical Center.

PUBLICATIONS

American Journal of Forensic Medicine and Pathology; *Journal of the American Medical Association*; *Nature*; *Science*; *New England Journal of Medicine*; *The Economist*; *The Merck Manual of Diagnosis and Therapy*; *The Book of Evidence* by Peter Achinstein (Oxford Univ. Press, 2001); *Fire Engineering Magazine*; *Nature Structural & Molecular Biology*; *Clarke's Analysis of Drugs and Poisons*; J. Weisberg, "This is Your Death" (*The New Republic*, July 1, 1991); *The Lancet*; *Accidents in North American Mountaineering*, 2007, edited by Jed Williamson (The American Alpine Club, Inc. 2007).

NEWS SERVICES

BBC Worldwide; CNN; *The New York Times*, PBS.